Telco Global Connect

TELCO GLOBAL CONNECT
Vol : 3

Author : Sadiq Malik

Copyright © 2016

ISBN-13:978-1523673513

DEDICATION

" For Dr Jean Grey...my Phoenix Soaring "

This book is dedicated to handling and resolving the mega challenges facing the Telco Industry even as the Internet is morphing from the global library into the global supercomputer. The book reframes the principal challenges through various technology , commercial and financial perspectives. This will help Telco professionals gain deeper insights to overcome burning issues and even profit at the end .

The Key Challenges are :

Telco Challenge #1 : The Quest for Business Model Innovation
Telco Challenge # 2 : The Quest for the Digital DNA
Telco Challenge # 3 : The Quest for Business Transformation
Telco Challenge # 4 : The Quest for Data Monetisation
Telco Challenge # 5 : The Quest for relevance in the OTT World
Telco Challenge # 6 : The Quest for Customer Loyalty
Telco Challange # 7 : The Quest for Softer networks
Telco Challenge# 8 : The Quest for Connected Machines
Telco Challenge # 9 : The Quest for Financial Acumen
Post Scriptum :   A peek into the Future

For network operators—pressured to increase ROI on existing network assets, and to become more agile and responsive to the changing needs of existing and prospective customers—a challenging business climate increases the need to develop an integrated perspective of the value chain and forging partnerships.

" It is not the strongest of the species that survives, nor the most intelligent that survives. It is the one that is most adaptable to change."

Charles Darwin

PREFACE

Mobile technology is redefining our lives and making it increasingly connected. From health and education to transportation and smarter cities, the proliferation of mobile communication and a connected life is now well established and here to stay. Most experts agree that the following   trends in digital technology will give rise to serious challenges and opportunities for Telco professionals in the next decade.

• A global, immersive, invisible, ambient networked computing environment built through the continued proliferation of smart sensors, cameras, software, databases, and massive data centers in a world-spanning information fabric known as the Internet of Things.

• "Augmented reality" enhancements to the real-world input that people perceive through the use of portable/wearable/implantable technologies.

• A continuing evolution of artificial intelligence-equipped tools allowing anyone to connect to a globe-spanning information network nearly anywhere, anytime.

• Disruption of business models established in the 20th century (most notably impacting finance, entertainment, publishers of all sorts, and education).

• Tagging, databasing, and intelligent analytical mapping of the physical and social realms.

Today, communications service providers deliver traditional and IP services that span voice, data, video, content, prepaid and postpaid, fixed and mobile. To succeed, service providers must change to an infrastructure that supports new business models, real-time customer interactions, and new partners and channels.

Driving this transformation are underlying business applications such as billing and revenue management (BRM), customer relationship management (CRM), and enterprise resource planning (ERP).

By the end of the next decade, almost every application or service we can imagine will be enhanced by the application of enormous computation enabling widespread applications of capabilities like mining, inference, recognition, sense-making, rendering modeling as well as proactive contextual computing,

Companies should strive to incorporate more agility and SOFT in their processes and IT systems, which will enable them to respond faster to changes in customer requirements and market conditions. Everyone involved in value creation must focus on innovation. New technologies, new products, and new services pave the way for penetrating new markets

Telco Challenge 1 - 2016 : The Quest for Business Model innovation !!

The arduous task of creating fresh and viable business models in the hyper competitive Internet era bedevils Telco CxO's and their financial backers. All markets are becoming mature and emerging market growth will slow in the next few years. Have Telcos ever considered the nightmare scenario when Apple and Amazon offer handsets plus e-SIMs. Contracts and billing could be handled via iTunes or Amazon accounts. The telco would thus become anonymous, a mere wholesaler of network capacity with no end-customer relationship of its own.

Most Telcos focus on matching and beating their rivals. As a result, their strategies tend to take on similar dimensions. What ensues is head-to-head competition based largely on incremental improvements in cost, quality, or both. So how do you leave rivals behind while sustaining spectacular growth for your company? Invent entirely new markets where no competitor has yet ventured. New business models are required where operators make money from new customers and ecosystem partners rather than exclusively from end-users.How can companies create breakthroughs in value and performance?

Telcos are being disrupted because the basis of competition in mobile has fundamentally changed from "reliability and scale of networks" to "choice and flexibility of services", representing transition from "mobile telephony" to "mobile computing". Telcos must move their innovation focus from technologies (be it HTML5, NFC, IMS, VoLTE, M2M or RCS-e) to ecosystems. That requires a much better understanding of how ecosystems are engineered, and how ecosystems absorb and amplify innovation.

"Communication networks are facing a lack of scalable and sustainable architecture to meet the challenges ahead in terms of data traffic increases, video uploads and downloads, and enhanced M2M communication. The network of the future has to be highly elastic in order to facilitate the adding or dropping of capacity and real-time provisioning of service. It needs to be highly orchestrated by key business imperatives, such as customer satisfaction, and it must be highly integrated so that synergies are fully embedded and captured across fixed and mobile, across borders and across segments."

(Bruno Jacobfeuerborn Deutsche Telekom CTO)

Companies that want to be successful in the current environment have to fundamentally scrutinize their business model on a regular basis and challenge its components if necessary. The overarching goal of a business model is to address a business opportunity in such a way that value is created for customers as well as for the company. A business model encompasses the addressed value potential, the customer interaction, as well as the value creation model.

Although many Telcos believe that they urgently need to build strong digital businesses, most are struggling to do so. Creating a Digital Telco means looking beyond traditional telco business models in the context of the changing telecom value network.The challenge for Telcos isn't that OTT companies outspend or " out imagine " them in digital innovation. It's that marginal cost analysis steers Telcos towards investments in capabilities that were relevant in the old basis of competition, rather than toward developing new capabilities relevant for the new basis of competition.

"I think you almost have two choices as an operator: either you say that you can be very, very efficient with pipes and that somebody else can deal with all of that and you're going to have the lowest cost and the best quality on my pipe, transmit as much data as possible "

(Hélène Barnekow TeliaSonera CCO)

The mobile industry is undergoing a dramatic rethinking of business foundations and supporting technologies. In many ways, technologies such as cloud, software-defined networking and 5G result in a "software is eating the network" end game. This in turn will promote opportunities that are much larger than just selling voice and data access meaning digital commerce, advertising, energy services, smart home , e-health ,M2M , Connected Car , Big Data , IoT etc .It is for the Telcos to adapt , improvise , transform to profit from these opportunities. The Telco that is best able to connect the trinity of technology, content and segment will be able to reap superior profits

" NFV and SDN concepts are at the core of our strategy. These help us realise our future network vision, which is a mutli-service, multi-tenant platform where we can respond more quickly and efficiently to our customers' needs. With NFV, we're able to dynamically reroute traffic and add capacity without adding new boxes. With SDN, we're removing pre-defined physical limits of the network by shifting control from hardware to software. These allow the network to become simpler, more scalable. They also allow us to reduce costs significantly and more quickly address customer needs "

(John Donovan, Senior VP– Technology and Network Operations, AT&T)

Perhaps one of the most successful new age " experience " players to date is SK Planet, which was set up in 2011 by SK Telecom, Korea's largest wireless operator, to offer multiple add-on experiences for both retail and business subscribers. They include MelOn, already Korea's largest music portal, with 17 million subscribers, has also been launched in Indonesia; 11st provides an e-commerce platform with related advertising and marketing intelligence services.It is now the country's second-largest e-commerce platform and largest player in mobile commerce; "T ad" is a mobile ad platform that enables personalized ads on mobile apps running on smartphones and tablets; "T map," a GPS-based navigation service platform with more than 10 million subscribers, also offers location-based services to businesses.

If you were to ask execs of France's incumbent Telcos who they loathe publicly and admire secretly they will probably mumble FREE and Xavier Niel. Now how did Free and Niel freak out fat cat telco execs who were relaxing on their oligopolistic sofa ?? In April 1999, Free entered the Internet service provider (ISP) market with a simple, no-subscription service. This commercial strategy was at first based solely on providing "Pay-as-you-go" access and enabled Free to win a large share of the dial-up market with relatively small advertising outlay as compared to its competitors. After completing the rollout of its telecommunications network and interconnecting with the France Telecom network in April 2001, Free was in a position to control the cost structure of an offering based on Internet connection time.

The Illiad Groups's network (Free is the brand name) enables it to design sustainable service offerings that are easy to understand, technically sophisticated and attractively priced. Free has capitalized on the different nuances of its brand name, transforming it from a name implying that the offering is free of charge into a name associated with high-quality paid services and the freedom offered to users of these services. The high speed broadband Internet access offerings are among the most competitively priced on the market in their respective segments while providing a high quality of service. This positioning is a central factor in the Group's strategy and is aimed at creating the right environment for lasting and profitable growth . Through the use of its network and by building on its experience in dial-up offerings, Free has developed a high-quality broadband access offering which is attractively priced and, where possible, makes the most of the opportunities afforded by the unbundling of the local loop. Today French broadband rates are among the lowest in the world.

After thoroughly disrupting the ISP market by introducing supercheap Internet access Free / Iliad pulled it off again with its mobile offering based on a 3 G license (2.1 GHz and 900 MHz bands.) With rates as low as €2 (about $2.60) a month, the Free Mobile phone service has lured an estimated 2 million landscape by using very clever technology, marketing and financial tricks. As a company with a hacker culture, Free is a good example of how to execute against well-established competitors. Free has created an offering that you cannot ignore. Imagine a mobile phone plan, with unlimited talk, unlimited SMS and MMS messages, tethering and, even more important, unlimited data with a speed reduction after 3 GB. Usually for that plan in the U.S., you would pay more than $100 for limited data with a two-year contract. In France, it costs $25 per month and there is no contract. Competitors had no choice but to lower their prices, even if it meant lower margins and lower infrastructure investments.

Operators need to realize that extending connectivity alone cannot keep them afloat. Instead they require software, device and service strategies that can add value and at the same time differentiate them from competition.In the future, the primary operational mode for large players might well be as aggregators of massive services. The key is to open the platform and gain as much partner power as possible. This is the fundamental reason why concepts like Web2.0 and P4P became important. NTT DoCoMo pursued the concept of Value Innovation : a new way of thinking about and executing strategy that results in the creation of a blue ocean and a break from the competition. More importantly, value innovation defies one of the most commonly accepted dogmas of competition-based strategy — the value-cost tradeoff.

NTT DoCoMo picked up its business model from Blue Ocean thinking . In a nutshell Blue ocean strategy is about creating and capturing uncontested market space, thereby making competition irrelevant. In more specific terms blue ocean strategy is about a whole system alignment of the value proposition (utility minus price) by creating an offer that dramatically raises buyer utility at the right price for the mass of the market; profit proposition (price minus cost) by creating a leap in value for the company itself by making a tidy profit; and people proposition by overcoming key organisational hurdles and building execution into strategy formulation. NTT DoCoMo is the first company to make money out of the mobile internet. In a very competitive industry engaged in a technology race and strong price erosion, NTT DoCoMo was able to achieve superior performance when it launched its novel i-mode services in February 1999. It was an immediate and explosive success in Japan. As with NTT DoCoMo, the goal for a firm's blue ocean strategic move is the pursuit of value innovation — a leap in value for buyers and company alike. This comes from simultaneous pursuit of differentiation and low cost.

More on Blue Ocean : (https://en.wikipedia.org/wiki/Blue_Ocean_Strategy

If telcos do their homework with respect to broadband access strategy, customer experience , transform themselves into a cost efficient lean operating model backboned on digital,they have a real fighting chance to recover the ground lost to the OTT players. In order to achieve its transformation strategy, KPN Netherlands mapped out four concrete objectives: fixed and mobile service convergence; full utilization of current network resources; considerable decline in CAPEX and OPEX; and delivering a variety of new services, which are built on an All-IP network. These services would include: multimedia personal communication services such as, voice, video, photo, data, message, and PTT; IP corporate communication such as, an IP private line, IP PBX, and multimedia service; and multimedia entertainment such as, games, IPTV, the worldwide Web, and Portal.

Through creative partnering and innovative risk sharing options, new managed services and outsourcing business model options provide the framework for creating a next generation enabled portfolio of services for consumers and enterprises ready for 2 sided business models. Instead of short-term tactical advantages, the focus is firmly on long-term strategic gains by identifying blue ocean market segments where competition becomes irrelevant . And only then will Telcos be able to reap the full benefits of managed services via trusted partnerships.

"Operators are] shrinking in relevance too fast. Many partnerships (with content and internet players) have been for loyalty, not for revenue streams. We need to find different ways of charging our customers rather than complain "

(Johan Dennelind TeliaSonera CEO)

Its time high time for Telco CxO's to think about newer business models with alternative revenue sources while earning little or nothing on basic voice / data. How about an ad funded business model pioneered by Blyk ?? Blyk was an ad-funded MVNO focused on the 16-24 year old market (although they positioned themselves as a 'media company'). It gifts minutes and texts to customers in exchange for the right to send advertisements to them. Users complete a set of questions about themselves when they sign up, giving Blyk information about their preferences. Advertisers market their products and services via text to Blyk users based on this profiling and Blyk got paid to deliver the advertisement. So, at first glance, Blyk reversed the normal revenue model for operators: it collected money upstream (advertisers) and paid out for delivering services to customers: in reality however Blyk was making money from customers as well (2 sided business model) via : Termination charges from off-net callers : Operators are both receivers of money from end users (when originating the call) and receivers of money from other operators (when terminating the call).

For the uninitiated 2 sided or multi sided models explanation can be found here :

https://en.wikipedia.org/wiki/Two-sided_market

So every time a Blyk user receives a call or text from an offnet customer the originating operator pays Blyk for termination. In turn, Blyk obviously paid some of this termination charge out to its network supplier (Orange) but one can hazard that the company still made some margin on this. Typically 16-24 year olds, like the rest of us, have a predetermined communications budget – "I will spend $x on my phone each month". The fact that Blyk gave users free calls and texts did not stop users from spending this money. Blyk's users would simply display the same behaviour that every Telco exec is familiar with: increased communications usage as the price reduces (price elasticity). Blyk offered 217 free minutes and 43 texts, so users will be profligate with their communications. They would use this free allowance up and STILL spend at least some of their previous budget.)

Two years after its launch , Blyk became part of France Telecom's Orange network in the U.K. Despite successes in luring major advertisers and producing effective campaigns, Blyk never reached its hoped-for scale. It topped out at 200,000 customers in its first year of existence.A co-founder of Blyk, admitted, "Advertisers want to see scale but we were not rolling out quick enough." Blyk ran more than 2,000 campaigns and attracted quality marketers to its service: Coca-Cola, Colgate, Penguin Books, L'Oreal, Lucozade and the BBC all partnered with Blyk to create cutting-edge mobile marketing aimed at an exclusively young audience. Many people in the industry view Blyk as a pioneer; a trailblazer for mobile marketing whose innovations have set the standard for the industry and will continue to play an important part in its evolution.

Perhaps the most obvious lesson for other operators is that there is value in 2-sided business model where the Telco earns both from upstream and downstream clients : in Blyk's case from advertisers and subscribers ! While Blyk struggled to make a return because of scale (or lack thereof) ,it did enough to show that for operators with large existing (youth) customer bases the ad-funded model could be fruitful. It was not mere marketing fluff that Blyk refers to itself as a media company rather than a MVNO. It showed that Blyk's management considered the advertising community as its primary market and end users as 'members' rather than customers. Indian carrier Aircel teamed with youth media firm Blyk to launch a content and advertising service in November 2010.Blyk still has its eye on the prize: developing the capabilities – in partnership with mobile operators – to be a game-changing engagement media in reach and response !!

The proliferation of digital media offers fertile ground for service provider-led innovations. Increasingly, SPs are employing a two-sided business model—focused on both business customers/partners and consumers—to monetize new digital media opportunities. On the business-to-consumer (B2C) side, this often means launching next-generation services that integrate web content with their traditional television or mobile services. Increasingly, these services are popular with consumers aged 30 or younger, and can positively differentiate service providers that offer them. Examples of these new services in the TV and mobile arena include user-generated videos, video-chat-enhanced social networking, and role-playing games.

AT&T is an example of a major service provider that has adopted this two-sided approach to innovative services. On the consumer side, the AT&T U-verse IPTV service utilizes a broad B2C value-added service offering that includes integration between the web and TV to compel users to upgrade to more advanced set-top boxes and higher-tier broadband plans:

• Web integration includes integration with a user's online Flickr photo-sharing account. While this capability is part of U-verse, it offers additional revenue opportunities from ad insertions.
• Device integration includes the ability to program the DVR from a user's mobile phone or personal computer; TV-based call management of fixed voice line service; and photo and music media sharing from a PC. Media sharing among devices is available only to U-verse subscribers, encouraging additional subscriptions from users who value this service.
• Interactive TV features are also part of the U-verse package, including greater interactivity; customizable personal screens showing weather, stocks, traffic, or other user-specified information; Yellow Pages directory; video on-demand, and games. Most of these features are included free as part of the U-verse plan, but there may be ad-supported revenue in the future. An exception is on-demand video, which is charged on a per-channel or per-movie basis.

On the business side, the Digital Media Solutions portfolio from AT&T helps digital media B2B customers such as web players and media companies manage, repurpose, and deliver content globally, delivering video and rich media on multiple screens and in multiple formats. Capabilities include:

• Intelligent content distribution services, offering web, file, and video acceleration via global CDN for both live and prerecorded on-demand video
• Broadcast video services, offering broadcasters, film studios, and production houses full-motion video transport services for video, news, and advertising content
• Ancillary services, such as content management, digital rights management, digital asset management, advertising, and transcoding. These are offered through partnerships with third-party solutions

Current Telco mainstream offerings to the Enterprise market are also based around capacity and hosted services, sometimes complemented by IT outsourcing projects. Mass-market consumers and Enterprise customers alike are increasingly demanding rich, portable, personalised, access and device independent services from their Telco Service Providers.Telco 2.0 and Web 2.0 components creates more value to the Enterprise . For example Telco resources can be embedded with the Enterprise applications to identify the real time location and distribution of a service engineer's customers (using Google Maps and Location feeds) to view the geography of the area covered.

Essentially in the 2 sided business model Operators can compete and partner with other players, such as the internet companies, in helping businesses and consumers transact over the internet. Networks are no longer confined to communications but are used to identify and market to prospects, complete transactions, make and receive payments and remittances, and care for customers. The knowledge that operators have about their customers coupled with their skills and assets in identity and authentication, payments, device management, customer care etc. mean that „the networks" can be „enablers" in digital transactions between third-parties – helping them to happen more efficiently and effectively.

According to the Telco 2.0 Group , collaboration between service providers, understanding the needs of upstream industries and end-users garnered are absolutely key to the creation of the two-sided business models:

1. Collaboration is necessary to create new markets for upstream and downstream customers. This is because an application that addresses only a proportion of customers belonging to only one network is significantly less likely to be attractive and successful than one that can be used by or address anyone (e.g. the impact of SMS interoperability on usage).
2. Understanding the needs of "upstream" industries is critical – to understand their value chains and the pain points that new services can valuably address. This requires insight which is outside of the most traditional Telco's expertise – deep into the processes of other business that are not necessarily obviously related to communications.
3. The needs of end-users are always important in service design, but in this instance there are two further twists. First, the needs again are not purely about communication and some are therefore outside of the traditional Telco world view. Second, some complex issues of security and privacy can be invoked by new uses of customer data, e.g.s Google Street View, Phorm. These issues can be addressed but not without clear understanding of the users' needs and wishes and a product / service design that addresses these needs effectively.

VisionMobile feel that , to grow Telcos could either...diversify, which means branch out into new markets by investing their profits into that new market, and try to turn that new market into a source of profits in its own rights; or...grow asymmetrically, which means branching out into a new market with the intention of not turning huge profits within that new market, but rather to drive profit in the core market. They have talked about Asymmetric business models

Asymmetric business models cross industries and force profits to migrate from one market to another.Google for example uses them to disrupt industry after industry: from mobile (Android), to television (Google TV and Chromecast), enterprise software (Google Apps), personal computers (Chromebook), travel (Google Flights), energy (Nest) and transportation (Android Auto and self-driving cars).The new generation of messaging apps have asymmetric business models, they can sustain free services indefinitely and forever change the dynamics of the mobile messaging market. WeChat, Viber, Line and others monetise by using their platform for e-commerce, selling digital goods (stickers, games), physical goods and services (like taxi rides). They don't have to charge for the messages or even voice calls to be commercially successful.

OTTs do not compete for telco service revenues; instead, they compete to control key links in the digital value chain, with business models that span consumer electronics, online advertising, software licensing, e-commerce and more. Thus, competition is not symmetrical, because unlike carriers, OTTs do not bear the burden of providing mobile Internet service. Connectivity may be as important to their business model as gas to a car; yet, it's the telcos who supply it, not the OTTs themselves. This asymmetry makes it difficult for telcos to protect the profitability of some legacy business model. Which bring us to Scenario Planning !!

Scenario planning, is a strategic planning method to devise flexible long-term plans. It is in large part an adaptation of classic methods used by military intelligence : WAR GAMES. At Royal Dutch/Shell it was viewed as changing mindsets about the exogenous part of the world, prior to formulating specific strategies. Scenarios provide a mature insightful way to lay out alternative futures to challenge cherished but perhaps irrelevant assumptions on the future of the high capex volatile industries like Oil and Telcos. The aim of scenario planning is not to identify one specific future, but to explore possible futures and to identify the mechanisms that drive the future in a particular direction.

Scenario-based strategic planning achieves its cognitive effects through a structured, tool-based approach .Here are some simple yet plausible scenarios for the Telco Industry !! Traditional strategy tools are designed for stable market environments, but fail predictably when applied to innovation under conditions of uncertainty and rapid change, which characterizes today's telecom market. Telcos can no longer afford the luxury of sluggishness, strategic inertia or blind spots.The good news is that various scenarios show an abundance of opportunities for Telcos.

In the Technology scenario Telcos obtain a competitive advantage by developing / delivering new proprietary technologies. At the same time customers' buying decisions are mainly influenced by the technological characteristics of the innovations. And consumers have become technology savvy and technology addicts. Under these circumstances, the most relevant strategy for telcos is to invent new technologies and push them to the market. In the long run this scenario may mean the disappearance of fixed networks as the leading technology for transporting voice and data. Even if physical networks remain, in this scenario they will play a minor role. They are what roads are to transport: they need to be there, they need to be maintained, but apart from that nobody really cares. The customer is interested more in cars (Smartphones), than in the type of asphalt (LTE 4G) he / she drives on.

For example Google is highly innovative, only distributes services online and has completely disrupted existing business models, putting pressure on such diverse industries as traditional software, newspapers and publishers. In the technology world, telecom companies that are disruptive innovators will do to the current telecom industry what Google has done to software. They will redefine existing industry boundaries completely, based on technology.

In the Commodity scenario there are no distinctive technologies that give telcos a competitive edge. Instead, there are a number of standard technologies offered by vendors. Consumers recognize that technological change is incremental : they are all very much aware of the standards that exist and have a pretty good idea about what they want. They do not need to go to a store to check the hardware, but just order it online. Telcos will have to offer the technologies and the connected services in a convenient and efficient way.

Gaining economies of scale to serve mass markets cheaply is the key to company survival here. The role of the Telco is best described as a standard direct packager: it takes standard technology and services from other companies, then packages them and offers them on the web with little thinking about product differentiation. For example Dell does not innovate its technology, but buys components from vendors. Next it ties those components together and ships the resulting product fast and cheaply to the consumer. It is a completely online business model. As most people now have an understanding of what they want from a pc or laptop, or they have a smart cousin who can help them articulate their wishes, there is no need for a store with staff explaining what the basic choices are.

In the Customised scenario each segment has its own demands in terms of products and services and how these are to be delivered. R&D focuses on serving this diversity of market segments. Telcos will follow a demand pull strategy in developing innovations. They will listen to the market, identify the needs of different consumer groups and innovate around this need. This may lead to different versions of products and services being offered in the market or to entirely different technologies being developed for different segments.

The role of the Telco is that of a market driven innovator which may lead in two directions. The first one is reactive: creating innovations based on a profound understanding of consumer needs. The other road is more proactive: creating innovations that create new market segments that other telecom companies have not yet thought of.

For example Apple innovates based on superior understanding of what consumers like. In doing so they deliver superior technology to existing markets (the iPod that replaced other MP3 players), but they also create new market segments by supporting online communities. Their approach is decidedly multichannel with much online business, but also investments in Apple stores and resellers.

In the Segmented scenario, telcos source innovation from vendors and package those standard technologies for specific market segments. In many ways this scenario is closest to the current strategies of most traditional telecom incumbents. They have decreased their investment in innovation and have embarked on a multichannel strategy. In the segmented world technology and content sourced from vendors need to be integrated and they need to be matched with the diverse market segments that exist.

The company that is best able to connect the trinity of technology, content and segment will be able to reap superior profits. However, a critique on this strategy is that it tries to be everything for everyone. There is little focus on either specific technologies or segments. The risk that niche players enter into the most profitable segments may be high. Have Telcos ever considered the nightmare scenario when Apple and Amazon offer handsets plus e-SIMs. Contracts and billing could be handled via iTunes or Amazon accounts. The telco would thus become anonymous, a mere wholesaler of network capacity with no end-customer relationship of its own.

Telecommunications companies are facing difficult times these days: their products are commoditised, competition is fierce and pressure on margins is high. Some operators with foresight are trying to set themselves apart through better service. Services, by being less visible and more labor dependent, are much more difficult to imitate, thus becoming a sustainable source of competitive advantage. Reaching out to and understanding the needs of customers, both current ones and those who may consider shifting from competitors in such turbulent and competitive era, is an important element of the service strategy. As such Telcos would do well to learn from other industry best practices to gain a competitive edge in and expand beyond their core markets.

Other industries, such as computer or telecom hardware, have shown that business development based on new services can be a successful road to new growth. In 1993, CEO Louis Gerstner initiated a transformation at IBM. The transformation involved a change from a hardware and software business to a solutions and services business and from a regionally aligned organization to a global organization. IBM committed itself to business and cultural change , invested in talent and the right financial and IT systems to support them and placed strategic bets on IT as a utility service and hosted storage. The transformation created opportunities for cost savings by encouraging development and use of enterprise-wide technology platforms.

Fortunately with the ever evolving technology and more complex products, incumbent operators have a powerful asset to leverage: their technical field service organization and capabilities. In order to capitalise on further strategic growth opportunities, Telcos should consider an option of developing a dedicated service organization which has control of its entire value chain, primarily focusing on multiproduct communications services for mass segments. This means a dedicated unit, focusing on developing the service business, having full control over its entire value chain, freedom and leeway to develop its business, full management attention and support to execute its mandate.

The primary focus of the new service organizations is service innovation excellence and ability to scale customer solutions for rapid growth across well defined customer segments based on their real communications needs. Customer front-end responsibility (marketing and sales) should be placed into new service Org. The consolidation of the service offering under a single division is normally accompanied by a strong initiative to improve the efficiency, quality and delivery time of the services provided, and the creation of additional services to supplement the basic offering. The consolidation of services also comes with the development of a monitoring system to assess the effectiveness and efficiency of the service delivery.

Transitioning from product manufacturer into service provider constitutes some managerial challenges. Services require new organizational principles, structures and processes. Not only are new capabilities, metrics and incentives needed, but also the emphasis of the business model changes from transaction- to relationship-based. Be warned that research has shown it takes a serious effort by senior managment to build the structures, capabilities, processes and systems to seize the service opportunities.Successful service companies do not start from scratch – they are built up on the basis of existing units and businesses with the best suited set of service assets and capabilities such as customer knowledge ; service development, standardization and roll-out capabilities for complete service delivery.

A focus on forward-looking IT investments (funded by reductions in maintenance costs for today's systems) will be essential to support the service organisation. Social media collaboration platforms support all service operations: enterprise case management, call centers, customer portals, websites, and integration with social media channels. A knowledge base provides answers to your agents and your customers through all your channels, increasing deflection rates and reducing time spent per case, keeping your customers happy and loyal. Cloud based CRM platforms can support the customer service team to improve the way they managed everyday customer interactions.

Communications service providers have a number of attributes that give them a potential marketplace advantage: an extensive customer base, distribution muscle and knowledge of customer preferences through CRM and billing systems. The opportunity is to become an integrated digital services provider across platforms and mobile devices—convincing customers that a communications service provider can effectively serve as the hub to meet their communication and entertainment needs. According to an Accenture survey, the areas that show particular promise include cloud services and location-based offers.

Cloud Customer Portals give customers a true online service experience, ensuring they have the flexibility to manage their interactions with their Telco entirely online if they chose to do so, and enabling customer service questions to be managed, just like an order coming in from a field sales team person. Customers can create orders online for new and replacement products including, phones, accessories, and SIM cards, and then track the status of the order through to shipment. When Sprint acquired Nextel, over 5,000 employees in over 1,100 retail stores and 800 dealer locations got busy collaborating on customer retention and churn avoidance. Tied together via an employee social network, disparate teams across both organizations focused their efforts to retain customers and build new loyalty programs.

Delivering good services as part of a core product offering does not suffice as the sole differentiator in highly competitive telecommunications markets. Investments in new radio access technology bring along radically new network economics leaving mobile operators with the quest to gear their network investments towards a cost optimal access, backhaul and core portfolio. It is critical to cut spending on low-value activities, and redeploy it to investments that generate growth, margins and true differentiation. Being able to accurately identify where value is generated at all levels of the organization – from divisions to specific products or offerings to particular customers – is a critical managerial competence.

Customer ownership and distribution power give communications service providers a strong foundation on which to build to meet consumers' ongoing communication and entertainment needs. Providers have an opportunity to improve their return on investment by monetizing better connectivity. They can also extend their partnerships across the digital ecosystem to provide a seamless customer experience. This will require deep insight into subscriber behaviors, new forms of collaboration within the industry, new capabilities within the organization and an ability to constantly innovate to keep pace with today's demanding consumers.

Perhaps one of the most successful new age " experience " players to date is SK Planet, which was set up in 2011 by SK Telecom, Korea's largest wireless operator, to offer multiple add-on experiences for both retail and business subscribers. They include MelOn, already Korea's largest music portal, with 17 million subscribers, has also been launched in Indonesia; 11st provides an e-commerce platform with related advertising and marketing intelligence services.It is now the country's second-largest e-commerce platform and largest player in mobile commerce; "T ad" is a mobile ad platform that enables personalized ads on mobile apps running on smartphones and tablets; "T map," a GPS-based navigation service platform with more than 10 million subscribers, also offers location-based services to businesses.

By consolidating a wide range of services under one roof, on top of its successful core wireless broadband business, SK Planet now offers perhaps one of the most compelling customer experiences of any operator worldwide.To provide customers with exactly the right products and services based on their actions and requests, " experience " players such as SK have to become fully responsive to the correct interpretations of their customers' behavior, often in real time. The ability, for instance, to offer access to medical information services could follow evidence of increased interest in healthcare. So investing in data analytics capabilities to respond to customer data is a must. "Big data" offers much promise in this area, but it will require considerable investment.

According to experts at Booz " experience play " Telcos de-emphasize their network activities and in some cases even carving out their entire access network infrastructure—both passive and active. They can share these costly assets with competitors via network-sharing agreements, and then differentiate themselves through innovative products and services. In contrast, some recent mobile and fixed entrants in Europe and the Middle East have focused on deploying their own network infrastructure in order to become connectivity or platform players. Consequently, they have minimized their investments in customer-facing infrastructure, relying on an online presence for sales and deploying only flagship stores to serve as their bricks-and-mortar channel.

Telecom operators should look for opportunities for growth by both assessing their markets (the market-back view) and evaluating their own current strengths (the capabilities-forward view). Both points of view are critical; if the two aren't considered together, the result will likely be a capabilities system poorly aligned with market opportunities. The capabilities forward view seeks to find distinctive internal capabilities that can be leveraged in any number of ways: to grow into adjacent markets, to build innovative new services, or to increase network speed and capacity.The market-back view turns outward for market opportunities that might arise from new technologies or from opportunities that competitors might be overlooking or are not coherently pursuing.The goal is to become coherent: to strike a balance so that the right product and service portfolio naturally thrives within a capabilities system consciously chosen implemented.

The new business models (which means monetising Web 2.0 services / Data) require relatively low levels of incremental capital investment so, although they generate lower EBITDA margins than existing services, they can generate substantial CROIC margins. Cash Returns On Invested Capital (CROIC) is a good measure of company performance because it demonstrates how much cash investors get back on the money they deploy in a business. It removes measures that can be open to interpretation or manipulation such as earnings, depreciation or amortisation. Telcos tend to focus on the existing capital-intensive business (which currently generates CROIC of around 6% for most operators) rather than investing in new business model areas which yield higher returns.

Telco professionals have already realised that unless they shift to new business models based on a refined customer understanding of customer expectations they will lose even more revenue to OTT players. Telcos need to embark on a series of such changes in order to ensure that they can build upon their successes in delivering telecoms services. In the end, they will need to ensure that they continuously challenge established models and notions on their role if they are to truly innovate their business model.

---♠--

Telco Challenge # 2 : The Quest for the elusive Digital DNA

Digital commerce is evolving fast, moving out of the home and the office and onto the street and into the store. The advent of mass-market smartphones with touchscreens, full Internet browsers and an array of feature-rich apps, is turning out to be a game changer that profoundly impacts the way in which people and businesses buy and sell. As they move around, many consumers are now using smartphones to access social, local and mobile (SoLoMo) digital services and make smarter purchase decisions. For most telcos, the best approach is to start with digital commerce, where they have the strongest strategic position, and then use the resulting data, customer relationships and trusted brand to expand into personal cloud services, which will require high levels of investment. This is essentially NTT DOCOMO's current strategy.

"Everything that can be digitized will be digitized. Everything that can be connected will be connected. Bridging the infrastructure and ecosystem to digitization is a mega trend we see in the second phase of the Internet "

(Tim Hottages , CEO Deutche Telecom)

Communications service providers have a number of attributes that give them a potential marketplace advantage in the Digital world of commerce and comms : an extensive customer base, distribution muscle and knowledge of customer preferences through CRM and billing systems. The opportunity is to become an integrated digital services provider across platforms and mobile devices—convincing customers that a communications service provider can effectively serve as the hub to meet their communication and entertainment needs.

According to an Accenture survey, the areas that show particular promise include cloud services and location-based offers. In May, 2014 Norway's Telenor announced a new Group strategy and Digital Unit in order to boost its responsiveness to customer needs. But there is one Telco that has taken the notion of Digital imperative to stratospheric levels : TELEFONICA.. my favourite Telco Titan !!

Telefónica has been rapidly transforming into a Digital Telco par excellence. They aim to provide the digital products and services which will help to improve the lives of our customers by leveraging the power of technology. This ranges from developing new technologies for consumers to communicate with friends and family through to helping businesses and governments address new opportunities, improve operations and increase efficiencies.

As of February 2014, Telefónica implemented a new organizational structure, one that is completely focused on clients and incorporates this digital offering as the main focus for commercial policies. The structure gives greater visibility to local operations, bringing them closer to the corporate decision-making centre, simplifying the global structure and strengthening the transverse areas to improve flexibility and agility in decision makings.

So that's a global operational revamp and a challenging new networks strategy — a lot of change within a large global organization (about 130,000 staff and 323 million revenue-generating connections). Within this framework, Telefónica has created the role of the Chief Commercial Digital Officer (CCDO), who is responsible for fostering revenue growth. On the cost side, the Company has strengthened the role of the Chief Global Resources Officer. Both Officers will report directly to the Chief Operating Officer (COO), as will the local business CEOs for Spain, Brazil, Germany and the United Kingdom, in addition to the Hispanic-America Unit.

The reorganization comes at a critical juncture for Telefónica, as the operator is also embarking on a radical evolution of its network infrastructure with the introduction of NFV and, ultimately, SDN capabilities. "capture gross savings of up to €1.5 billion "in the next year", as well as having responsibility for "the synergies plan in Germany.Value-added services are the competitive edge, but the strategy – and how deep into the water a Telco goes in placing itself at the centre of the Telco 2.0 environment is very important as well. While other companies take a "defensive stance" focusing just on value-added services, Telefónica has fully embraced the multi-sided business model, acting as a middleman between end customers and other third party companies, according to a study by Telco 2.0 Research.

What are the technologies that will help communications service providers such as Telefonica , SingTel , Telenor etc meet current needs and the challenges of the future? According to Gartner, CSPs are actively considering deployment of a new breed of IT-centric services that involve the use of data, analytics and digital content. CIOs will need to transform business-only, support IT to effectively deliver customer-facing digital services.To transform CSPs into diversified service providers, Gartner said that CIOs must harvest business value from existing networks, IT and information assets, and hunt for new opportunities using digital services.To drive new revenue, CIO offices must acquire new skill sets to align IT to business strategies.

"The industry focus in the coming years will be on personal data, connected living, digital commerce and networks. To realise the necessary developments in these areas, "huge investments of $1.7 trillion need to be done from now until 2020",

(Jon Fredrik Baksaas,CEO of Telenor and chairman of the GSMA)

The priorities of telecom CIOs are related to business applications, BI ,analytics, customer service support systems, centralizing the billing systems and the convergence of Internet protocol. While CIOs work to deliver applications and innovation, they are also asked to be as cost effective as possible. Looking ahead, IT leaders from communications service providers need to be aware of the shifts brought by new technologies – and they will be requested to provide insights on how carriers should provide strategic plans and new services to be launched based on them. This may involve changes to the infrastructure layer to fit new demands.

You recall ofcourse the seminal IBM Whitepaper Telco 2015 that painted4 future scenarios for Telco players. One of the scenarios is called " Generative Bazaar ". In this scenario pervasive, affordable, open connectivity is enabled for a person, device or object, unleashing a wave of generative innovation. A co-operative of horizontally integrated network infrastructure providers (Net Co-op) emerges, based on catering to the needs of a multitude of asset-light service providers that package connectivity with completely new services and revenue models.

IBM modeling of its four scenarios suggests Generative Bazaar as the most attractive outcome in terms of revenue, profitability and cash flow projections. This is precisely what Telefonica's Digital dream is all about.Telefonica takes an entrepreneurial approach to identify and accelerate the latest technology trends, incubating new disruptive digital businesses (Wayra and Amerigo). They aim to deliver end-to-end digital products and services in B2B, B2C and B2B2C in a number of domains: future communications, machine-to-machine, financial services, media services, cloud computing and information security. They also develop the right partnerships so they can be the single source of the best digital experiences for its customers.

Telefonica believe its strategy is supported by "four pillars in the short term": growing revenue by extending the commercial offering to new services in the digital world; modernising networks and systems, through accelerating the deployment of the most modern technologies; increasing efficiency through simplification and cost-cutting as well as ongoing financial discipline, prioritising investment in growth projects that generate added value; and strengthening its leadership in the digital ecosystem, "by driving a new public positioning enabling the hypersector to re-establish balance in the value chain".

In essence Telefonica has reorganised itself to create , catalyse and capitalize on opportunities along the following vectors : • Innovation & New Business – focus on discovering and incubating the next generation of digital services. They take an innovative 'venture capital'-like approach to innovation, as well as identifying the right start-ups to work with, invest in or potentially acquire. Their crowd data offering, Smart Steps, sits within this unit.

• Digital Services – delivering high quality, integrated product offerings across a number of more established digital service areas, including Machine-to-Machine, Cloud Computing, eHealth, Financial Services, Advertising and Information Security. It also brings together all of Telefónica's capabilities in media services ranging from IPTV and satellite TV to over-the-top video, ensuring that Telefonica is a leading video company. •

Communications – enabling Telefónica to continue innovating in its core business of communications, leveraging its world-class expertise in IP and video communications.

Customer ownership and distribution power give CSP's a strong foundation on which to build to meet consumers' ongoing communication and entertainment needs. Providers have an opportunity to improve their return on investment by monetizing better connectivity. They can also extend their partnerships across the digital ecosystem to provide a seamless customer experience.

This will require deep insight into subscriber behaviors, new forms of collaboration within the industry, new capabilities within the organization and an ability to constantly innovate to keep pace with today's demanding consumers. They must establish value proposition for third-party providers, including interfaces to network capabilities, service enablers based on open standards, access to ecosystem of partners, a commercial model and infrastructure support for common business process services, e.g.: self-service, e-commerce, billing Leaner operating models and new leadership roles designed to spearhead digital growth are a critical success factor for Telcos, particularly those with large footprints incorporating markets at different stages of maturity.

Success in the IBM Generative Bazaar scenario is dependent on the provider's ability to achieve structural industry separation; a pervasive open network access infrastructure; support for third-party application / services innovation; a dynamic business design; and the ability to leverage advanced customer and network analytics.

By 2016, the fastest-growing markets will add nearly 1 billion new mobile connections and account for 56 percent of all mobile connections worldwide. While these hypergrowth markets will be the global driver, they also pose challenges that go beyond cross-border execution and cultural conflicts. CSPs will need to transform themselves into leaner, more industrialized digital companies. And Telefonica is the undisputed global leader in the digital endeavour.

Their Telenor Digital Unit has the challenging remit of developing services that will guide Telenor towards a future as an internet telco.Telenor Digital creates globally scalable solutions within next-generation communication services, cloud services, e-commerce, and the "Internet of Everything". Telenor Digital also enables global distribution of its own and third-party services and support new ventures within digital entrepreneurship.

Telenor believes that key to building a successful digital service is making it easy to use. Operators are ideally placed to securely remove obstacles to logging into an app or a web-page. By succeeding in the digital identity space, mobile operators will become a more visible part of consumers' everyday digital life. In response to this challenge the Telenor Global Backend is a common cloud¬-based infrastructure was developed to provide a global, shared system for giving Telenor's customers access to Internet service. This implies providing easy Sign Up and Log In – as well as frictionless payment.

As of today – nine out of 13 business units can offer payment of digital goods through Global Backend – e.g. on Google Play. In addition, six out of 13 Telenor BUs can offer services bundled together with a mobile subscription to their customers through the Global Backend infrastructure.

This also means that Telenor becomes more attractive towards partners because they have one integration point through the Global Backend, through which they can reach 172 million customers – instead of approaching 13 Business Units individually. At the heart of the global backend business model, is the customer ID – the unique key which all customer data is gathered around. The Connect ID is Telenor's global solution to authenticate end-users. In practice – it's a solution for signing up and logging into a service. In short, Connect ID offers easy access to all their services.

The Innovation City at Mobile World Congress (MWC Barcelona) is all about experiencing and exploring the latest mobile technology in the context of an extensive virtual cityscape, where amenities, services and businesses are transformed and given added functionality through mobile connectivity. At the City, you can see a demonstration of an International Coupon System which allows purchases to be made across the globe with a smartphone, via a one-tap, pay-and-redeem digital wallet which is being run by the GSMA Digital Commerce programme.

The new, digitally connected world is driving transformation, bringing with it new players, advanced applications, broadband services and higher QoS demands. Seizing this transformation by making your business and technology evolution work together is the key to profiting from market changes.BT has undergone a major transformation and continues to change. What was a traditional networks-focused, telco R&D organisation is now a 'softco', centred on developing software and using software development methodologies and practices. At the same time, networks and computing are quickly converging into what we know as cloud computing. Not surprisingly, BT's continuing transformation is now addressing the cloud services market.

Recent research by (JDSU / STL) has revealed an US$11Bn global opportunity for operators to monetize the data in their networks about places and people. The study concluded that demand for what it calls location insight services (LIS) will be driven predominantly by retailers that want to know more about local market trends and benchmark themselves against their competitors. Telcos are uniquely positioned to capitalise on LIS, as opposed to location-based services (LBS), which is proving more lucrative for over-the-top (OTT) service providers than telcos.

LIS is an extension of existing software and analytics systems although data collected by these systems requires additional processing before it can be re-packaged into something marketable.This information can be shared with external systems and can be integrated with data warehouses using cost effective techniques. In many cases the intelligence can be directly used with business intelligence solutions. While commonly available cell level location enables some of the use cases, building level location intelligence from a carrier grade LIS system significantly increases the value.

For example , LIS platforms can enable mobile operators to share precise location data with transport infrastructure planners to help the understanding of where in the transport network heavy traffic occurs and when. This insight can be used to plan effective investment in infrastructure, and increase citizen satisfaction by improving transport network efficiency. LIS platforms provide the trend insight about which venues receive the best audience attendances given certain parameters, which can then be used to create a framework for predictive audience modelling. This enables event planners to more accurately assess the viability of venue locations, without needing to carry out time and resource intensive customer research.

Some Tier 1 Telcos have recognized the opportunity and publicly made noises about providing this insight. Last year Telefonica Digital unveiled a new division called Telefonica Dynamic Insights, which is tasked with monetising its vast data resources. Their first product, 'Smart Steps', will use fully anonymised and aggregated mobile network data to enable companies and public sector organisations to measure, compare, and understand what factors influence the number of people visiting a location at any time.

These insights will help retailers tailor local offerings for existing stores and determine the best locations and most appropriate formats for new stores. A number of retailers are already helping with product development by providing user feedback. Smart Steps will also be able to help town councils measure how many more people visit their high street after the introduction of free car parking, farmers markets, or late night shopping.

SingTel is also attempting to buck the trend of telcos becoming just a big, fat dumb pipe that only competes on price. Their vision is to really go after the heart and soul of the digital consumer, ultimately to drive a deeper connection that substantially increases the value of SingTel.

Recently SingTel decided to restructure its business into three units : group consumer, group digital life and group ICT-in order to sustain growth, competitiveness, and innovation. With the reorganization, SingTel plan to reinvent its core carriage business, create and drive new growth platforms that leverage and strengthen the core, and turbo-charge its regional capabilities in ICT services. They broke new ground with the introduction of PowerON Compute Service. This state-of-the-art cloud solution provides enterprises with the business agility and cost effectiveness of public clouds without compromising on portability, compatibility, security and control demanded by enterprise IT organisations.

Companies like SingTel go beyond access business, positioning themselves as service providers and complementing their traditional sales and network operations with a third element, a "telco innovation factory" charged with developing and marketing new services. The innovation factory consist of access-centric services that use the existing network and IT platforms – in the e-health segment, for instance – or regional OTT-related offerings such as TV. Such services will be embedded in partners' service suites or, depending on the extent to which the ad valorem aspect is to be emphasized, on proprietary platforms that integrate third-party services.

One of the few countries in the world that punches way outside its weight class is Holland aka the Netherlands.The 16 million Dutch people pump out a whopping GDP that is almost 3 times South Africa with its 53 million people. Unlike resource rich South Africa , Holland has nothing except the diligence and intelligence of its people. From an early age there's a feeling instilled in the Dutch that they can do anything. You can cycle anywhere in the land without fear of being laughed off the road or run down by speeding maniacs.

Ofcourse Holland is not all cheese and tulips and its great people who gave military funerals to all the victims of MH17 flight tragedy over Ukraine.In Europe, the Dutch are among the frontrunners in the area of Digital Infrastructure (Internet connectivity, colocation housing and hosting).In many ways this infrastructure fulfills a gateway function, similar to that of Schiphol Airport and the Rotterdam Harbor. Amsterdam based AMS-IX remains the largest Internet exchange point in the world and the colocation housing market, centered around Amsterdam, produces strong growth rates. The Dutch also rank among Europe's elite in hosting. The services that encompass Digital Infrastucture on core Internet on one hand and colocation, hosting and IaaS on the other include :

• Internet Exchange: Parties that facilitate networks to interconnect with each other to exchange Internet traffic mutually (peering). This is typically done without charging for the traffic
• Transit Provider: Parties that provide network traffic in the 'core' Internet and connect smaller Internet service providers (ISPs) to the larger Internet
• Colocation: Delivering facilities (floor space, power, cooling, network connectivity) to enterprises and service providers for housing servers, storage and other computer equipment as an alternative for an in-company data centre
• Dedicated hosting: Delivering computing power and storage via equipment dedicated to a specific client but managed by the hosting provider
• Shared hosting: Delivering computing power and storage by sharing the resources of physical equipment among multiple customers
• IaaS: Infrastructure-as-a-Service, delivering computing resources (e.g. servers, storage) according to a model that meets the essential characteristics of Cloud computing: on-demand self-service by the customer, measured service (pay-per-use), rapid elasticity (any quantity at any time), resource pooling (multi-tenant model) and broad network access (infrastructure is available over the network via standardised mechanisms

The Amsterdam Internet Exchange (AMS-IX) is the largest in terms of connected Autonomous System Numbers (ASN). The significance of an Internet Exchange is measured by the number of peering networks (Autonomous System Numbers) and and the Peak Internet traffic in Gigabit per second. AMS-IX is a mainport for Internet traffic more than Rotterdam and Schiphol are for containers and passengers respectively. London, Frankfurt, Paris and Amsterdam form the leading group of colocation data centres hot spots in Europe. Measured in colocation supply m2 per € bn GDP, Amsterdam exceeds all other cities.As a result Netherlands is hosting the top of the world's technology and Internet companies as gateway to Europe and the Internet such as : Facebook , Twitter , Netflix , Akamai , Amazon , Google and on and on. For this reason world's largest service providers and e-commerce companies have chosen Amsterdam as their #1 or #2 Internet Exchange position in the EU.

Large investments in data centres within the Netherlands by corporate multinationals like Google and IBM generate additional employment. Direct employment in the Digital Infrastructure sector adds up to 7,600 FTE, of which 90% in the hosting sector and 10% in capital intensive housing. Operational expenditures and investments in the housing and hosting sectors drive indirect and induced effects to create additional jobs. Combined effects for the Digital Infrastructure add up to 19,000 jobs in 2013 with a projected growth of 8% a year.The real value of the Digital Infrastructure sector, however, lies in its significant impact on the much larger Internet economy and broader digital society.

A continues interaction between Digital Infrastructure, service innovation and online usage drives growth in the online ecosystem. Digital Infrastructure is part of a much larger online ecosystem generating at least ~ €39 bn in revenue in the Dutch economy. Including private investments, government spending and trade, the Internet economy in Netherlands adds an estimated €34 bn to the GDP which is approximately 5.3% of the total GDP and steadily growing. There is a strong correlation between the Digital Infrastructure and e-commerce which shows that the former is a key enabler because E-commerce application are hosted in data centres and e-commerce traffic flows over the Internet exchange(s).

The employment generated by e-commerce in Netherlands is estimated between 100,000 and 140,000. SaaS and PaaS are two of the Digital Infrastructure's closest relatives, generating 5,700 jobs in the Dutch economy. Google has invested €600 million on a data centre located in Delfzijl, the Netherland.The estimated additional employment that the data centre will provide is 150 FTE from operations and a 1000 FTE at the peak of construction. The presence of most major global data centre providers in the Netherlands is prove of the country's attractiveness in terms of Internet Connectivity, availability of required electricity capacity , economic and political stability and highly-educated and multilingual workforce.

A large and complex ecosystem of companies and other entities compete, collaborate and cooperate to construct and maintain the interconnected network of networks that is the internet. The ecosystem works, and anyone can download a web page or video, or activate a mobile app, because of common standards and a shared understanding among participants of the benefits of a vibrant and growing economic system. Countries need energetic digital service sectors. They are drivers of social and economic development, job creators, talent magnets and the exports of the future. Robust digital service sectors depend on a complex ecosystem that includes adequate infrastructure and an investment-friendly business environment.

BCG Reports that Emerging market consumers are embracing the mobile web as much more than a purveyor of convenience; they are using it to improve their well-being, intellect and earning ability. However the lack of broadband penetration in emerging countries – especially fixed, but also mobile – is a serious impediment to accruing the benefits of a first class Digital Infrastructure such as the one in Netherlands. This ought to represent an opportunity – many emerging markets are free to adopt new technologies, such as LTE and fibre, without the burden of managing legacy infrastructures. Progress has often been slow, however.

India, for example, has struggled to develop digital infrastructure. Fixed broadband reaches less than 10% of households, and while mobile penetration has hovered around 75%, it is dominated by 2G networks; 3G and 4G penetration is less than 5%. There is also a strong urban-rural divide, with mobile penetration in urban areas topping 160% while in rural areas it does not reach 40%.Indian mobile operators struggle with fierce competition, low consumer spending power and poor spectrum management. In Africa the mirror situation is even more pathetic !!

In the digital era, connectivity counts. It is impossible to imagine the country, sector, industry or area of endeavour that cannot benefit from digital services. The services enabled by digital technology are economic growth drivers, job creators, talent magnets and big sources of exports. The economics of many emerging economies make infrastructure (as well as other) investment tough. At the same time, a growing number of governments, companies and organizations recognize the benefits of expanding internet access as widely as possible. They also see that gaining access can have an outsized impact for people who live in particularly poor and remote areas.

Bridging this divide may require non-traditional and innovative approaches. Internet.org is a partnership started by a group of major technology companies (the founding partners include Ericsson, Facebook, Mediatek, Nokia, Opera Software, Qualcomm and Samsung) with the goal of working with governments and NGOs to bring basic internet services to people who do not have them. The underlying philosophy is that demonstrating the internet's value for free will cause users to want to pay for more or better services down the road (which is not too far removed from how internet use evolved in the rest of the world).

The policy-makers of the future (especially Developing countries) must be able to tackle the challenges posed by the digital economy. They need to consider the impact of policies on the entire value chain, including telecommunications, digital services and media, and ensure that any regulations that are deemed necessary are applied with a light touch and restraint. Perhaps most importantly, policy-makers need to take into account how quickly technologies and the innovations they enable are evolving.

According to GSMA "Beyond connectivity, mobile operators will play a crucial role in working together with a range of industry partners in health, automotive, education, smart cities and a range of vertical industries to accelerate the launch of valuable connected services," Unfortunately without continued investment and growth in mobile networks (especially LTE) and the deployment of multiple connected devices, the socioeconomic benefits of the connected life will not be fulfilled.

mHealth programmes are currently one of the most cost-effective ways of providing remote living assistance to aging and chronically ill patients. mHealth programmes provide faster response times, integrated record access and considerable ease of use to patients. Remote consultation and support is expected to address the growing chronic disease management issue by reducing the need for hospitalisation. Proactive mobile based care for patients with sudden health incidents can reduce the number of primary and emergency visits by 10%. Mobile technology can also be used for home monitoring, thereby reducing the need for face-to-face consultations.

In developed countries mobile interventions could help cut healthcare costs by 400 billion USD in 2017, help retain 1.8 million students in the education system, save one in nine lives lost in road accidents, and reduce CO_2 emissions by 27 million tonnes annually. Similarly in developing markets, mobile interventions could help save over a million lives in Sub-Saharan Africa, provide education access to 180 million students, save 25 million tonnes of food and encourage over 20 million commuters to start using public transport. (GSMA Connected Living Program)

Last year in Barcelona MWC , Telcos had the opportunity to generate value beyond basic connectivity through managed connectivity, stewardship services and platform innovation. The GSMA area was filled with interactive demonstrations of the connected life including the Aston Martin One-77, the bike of the future. It is fully connected and tuned into its own performance as well as the rider, including mobile health monitoring and electronics that track the bike's performance in relation to its environment.

There was also the Mantarobot showing virtual teaching through augmented reality and virtual presence and the Cooltra Connected Electric Scooter, the latest in smart city transportation, the GO! S3.4 from GOVECS which lets customers know when and where scooters are available and can be started with a phone via NFC. The widespread penetration of mobile networks offers a powerful platform to improve access to relevant content.

mEducation solutions already allow thousands of students in China, Bangladesh, South Korea and Indonesia to access course content through SMS and audio lessons. An mLearning student saves 86.7% of the cost spent by students taking the same training in a traditional classroom. Much of this is due to the elimination of the cost and inconvenience of travelling to attend courses. Inexpensive personal learning devices like the 35 USD tablet launched in India are further improving access to mEducation.

Mobile networks play a pivotal role in the development of the connected life providing a scalable, standardised global platform to support the growing demand for intelligent, secure connectivity.Examples of valuable connected services were amply demonstrated by leading edge Telcos at the GSMA Connected City showcase in Barcelona included :

In the Connected Home AT&T showed how people can use their smartphones and tablets to manage their energy, automate appliances and secure their homes through AT&T Digital Life. General Motors demonstrated how AT&T's 4G LTE network will transform the driving experience by enhancing safety, security, diagnostic and infotainment in the vehicles starting next year.

Deutsche Telekom, in conjunction with IBM, bought to life Smarter Cities for the Future using machine-2-machine technology to optimise urban services such as public transport, parking, energy, security and water management. Together with SAP, Deutsche Telekom also showcased Connected Port Solutions designed to optimise both road and sea traffic control as well as logistics and terminal operations in order to make port processes more efficient allowing larger quantities of goods to be trans-shipped in the port area.

Korea Telecom featured technologies that make our lives better including edutainment robots, automatic content recognition, smart home phones, a controlled motorcycle, eco food bins and cloud CCTV.There were also Smart Apps in your hand showing how we can live smarter with intelligent and unique applications including mobile K-pop music, integrated mobile payment and self-created M-learning solution.

Vodafone were showcasing their Smart Home, Smart City and Smart Mobility solutions.Smart Home illustrated how M2M technology can provide premium security services, enable remote health monitoring and even open and close doors remotely.Smart City demonstrated how Vodafone's Energy Data Management (EDM) solution, solar energy production monitoring, remotely controlled street lighting and digital signage are enabling the smart city.

Vodafone's Connected Cabinet solution demonstrated retail display cabinets that report on location, operational status and stock levels in real-time. Smart Mobility showed how M2M is transforming the automotive and transportation industries, be it through real-time information systems for public transport, enhanced drivers' experience with telematics services or through usage based insurance services with Vodafone Vehicle Connect.

Augmented Reality (AR) is a hot topic right now, attracting much of the hype that was reserved for apps a few years ago. Is AR simply the next stage of development for existing value propositions, or will it bring with it entirely new propositions that offer new revenue streams for the Telecoms ecosystem ? Augmented reality relies heavily on video display and database transaction technologies that will necessitate significantly more bandwidth than can be delivered by today's 3G networks. When this is the case it is easy to see that the current mobile network infrastructure will not support broad rollouts of augmented reality applications, which can require multiple megabits of bandwidth as well as universal coverage. Aahaa : enter 4G LTE with Digital Dividend Spectrum and you catch my drift !!!

In simple terms, 'Augmented Reality' applications and technologies bring users information that exists in the digital world and presents it automatically and intuitively in association with things in the real, or physical, world. Often, but not always, this information is from the web. AR is about creating, making explicit and displaying the relationships between the real and virtual worlds.

AR apps combine reality with a digital overlay allowing consumers to virtually try on glasses or items of clothing using their mobile phone. Usage of AR in the retail area can enable retailers to bring an internet-like experience into their stores, allowing consumers to see more information on a product simply by pointing their camera at it. With image-recognition transferred to the cloud, the number of images that can be identified will increase dramatically enabling retail brands to develop apps for use in-store. Usage by entertainment brands will also drive usage by driving consumers to download AR apps and try them out with products that they are familiar with.

AR has been studied in one dimension or another in labs for over 20 years. The most important drivers in AR uptake have been the release of new sensor-laden and processor rich smartphones with touch interfaces connected to cloud services by faster networks. In addition to enhancements in computing power, devices are increasingly pre-packed with GPS, compass, accelerometers and gyroscopes, adding to the now ubiquitous cameras and microphones. Thermometers, RFID and other wireless sensors are also appearing. At the same time a critical amount of information is available in digital format, meaning the potential for bringing the real and virtual worlds together through a mobile device is huge.

With the commercialization of high speed data network such as the Fourth Generation (4G) cellular networks via Long Term Evolution (LTE), the use of AR applications in healthcare represents a particularly compelling value proposition for cost reduction and of course saving lives. Many AR applications in healthcare provide the benefit of visualizing three dimensional data captured from non-invasive sensors. Applications range from remote 3D image analysis to advanced telesurgery.

Telcos are entering the market as application or software developers, or promoting the services developed by a third party. For example, Bouygues Telecom in France released the first in-house operator-developed mobile AR look up service in November 2009 with over 900,000 unique points of interest (POI), while Telefonica has a group in Barcelona R&D which is working on its own visual search technology. NTT DoCoMo offers its smartphone subscribers the intuitive navigation services "chokkan nabi" developed under contract for DoCoMo devices. Orange UK launched a free iPhone app and AR service for the Glastonbury Festival and others have released similar apps around special events.

Philippine telco giant Globe's Augmented Reality (AR) Christmas ad ran in major broadsheets with a colourful 5-part 'Parol' (Christmas lantern) depicting how Filipinos celebrate Christmas in diversity.The interactive ad allowed the audience to make their own Christmas "parol" and provided instant access for them to share it with their friends in social media. With AR, Globe hopes to establish a more personal engagement and intimate affiliations with brands, immersing them through different senses and forge a more robust interaction with its products and services.

Research firm Gartner Inc. proposes that augmented reality is one of the Top 10 strategic IT technologies of our time. According to research firm SEMICO, the total global revenue from augmented reality will touch 600 billion $ by the year 2016.They predict that more than 864 million mobile devices will be equipped with augmented reality by 2014 and more than 2.5 billion mobile augmented reality applications will be downloaded five years from now. Science fiction or not that's big business : so clever Telcos will stake a claim to this new digital market NOW by partnering with AR app developers , Advertisers , Retailers to create and monetise new AR apps.

Have you heard about Telco Accelerator Programs to accelerate Digital innovation using third party expertise. Basically when Telco operators establish new incubators as part of digital transformation initiatives with the creation of new business units to nurture entrepreneurial start ups.Operators are working with partners as part of accelerator initiatives, underlining how service innovation is increasingly reliant on harnessing expertise from across the telco ecosystem. Over the last few corporations have moved to an open innovation model or outsourced R+D. They're doing less basic research in house and essentially looking to bring that in through acquisitions. Ask Google and Facebook what they did to become so big !!

So why the sudden surge investment in Accelerators by Telcos ? For one the appearance of incubators in emerging markets underlines the role operators can play as leaders of technology ecosystems, particularly in countries where entrepreneurs may lack support structures. Accelerator programs offer a number of established advantages to operators — from lowering the risks and costs associated with new projects to enabling operators to tap into a wider talent pool. Not bad at all because it is not a merely a CSR initiative but a genuine effort to create new digital products and services with the help of small time developers on a shared risk basis : you know ...the same kind of small timers who built Facebook , Instagram , Twitter , WhatsApp etc into multi billion dollar titans that threaten to render Telco networks into unprofitable dumb pipes !! With the cost of developing new technologies coming down so dramatically, it makes sense for corporations to take smaller bets on new technology offerings developed in partnership with entrepreneurial apps developers .

As an example Wayra is Telefonica's Accelerator program. With the financial backing of Telefónica, and with the support of a global network of mentors, investors and partners, they aim to help the best entrepreneurs to grow and build successful businesses.They know the next digital revolution can emerge from anywhere, in technology nobody has the final word. If you have a business or an idea that uses technology to solve the problems of the future, Wayra is your place to make it grow so they opine. Their irresistible value proposition is : Our acceleration programme will give you funding up to $50,000, an incredible place to work, mentors, business partners, access to a global network of talent and the opportunity to reach millions of Telefónica customers. (http://wayra.co/)

In a typical deal in exchange for an equity stake of around 10 percent, successful applicants will gain access to an office space in the newly established Wayra Academy, financing, mentoring, access to technology expertise within Telefónica. After six months in the programme, they will be turfed out of their luxury abode and released to fend for themselves in the real world. Wayra is aimed at early stage technology startups less than two years old working in spaces including cloud services, financial services, M2M, digital security, e-health, mobile applications, social networks and e-learning.

A truly disruptive hardware or core technology investment can generate some lofty returns for Telcos who are battling with stagnating returns in the 4 G data era.Marrying incubation activities with a wider program of crowdsourcing, ecosystem collaboration and venture capital funding in an effective way is no easy task,and regular reviews of innovation agendas are a must. While many operators are focused on partnering frameworks and incubating new services as a route to service innovation, a robust internal collaboration structure should never be overlooked.

The « Orange Fab » is open to all US-based start-ups and is managed by Orange Silicon Valley, Orange's development center in San Francisco, California. Orange Fab aims to help start-ups grow their businesses and develop amazing products and services that benefit Orange and its customers. There were 4 - 6 start-ups selected to be part of the program that started in May 2013. Start-ups joining the program have access to mentoring sessions from notable Silicon Valley entrepreneurs, world-class engineers and experienced designers. All the start-ups that are selected for the program were offered $20,000 in funding. (http://orangefab.com/)

The French operator has expanded the Orange Fab program to markets including France, Japan and Poland, with additional innovation investments in the likes of India and Tunisia.17 Africa is proving to be an important focus area: in May, Orange set up an incubator called CIPMEN to support SMEs in Niger, having already set up incubators in Senegal and Mauritius. Meanwhile, Millicom has set up a tech incubator in Rwanda, launching a purpose-built facility in the capital, Kigali, in March.

Believe it or not one of the pioneering Accelerator programs (MAGNET) was initiated by Motorola over a decade ago. The MAGNet program provided application developers, included a membership structure, providing to the developer community a combination of products, services and benefits to produce robust and exciting applications and bring the best of them to market. The program included: Software Development Kits (SDKs) and tools, application testing and certification, technical support (both onsite and online), training, developer services, networking events and an interactive web-based community/portal to support developers worldwide.Applications that were marketed through the MAGNet programme were tested by independent test agencies against predefined criteria and evaluated as best in class. Motorola endorsed this approach as a way of ensuring high quality, safe and reliable software products in the increasingly open world of wireless application development. The company that birthed the cellular / mobile industry was a great visionary indeed !!

SK Telecom's venture capital arm launched SKTA Innovation Accelerator with a view to seed start-ups in core technologies such as data center and IT, in what the operator cloud technologies and software.16 Partnership is a key focus of the initiative, with SK Telecom Americas pairing entrepreneurs with strategic advisors as part of its value-add. Meanwhile, in July, Japan's KDDI announced the launch of its latest US$50m fund, Open Innovation II. At the same time, it announced that it had invested US$8m in four US start-ups — an educational social network (Edomo), a free digital publishing platform (ISSUU), a seat upgrading service (Pogoseat) and technology news media (VentureBeat). Fellow Japanese operator NTT DOCOMO already has a US$109m fund in place, while Softbank created a US$250m fund in 2013.(http://sktainnopartners.com/overview/)

In February last year , Telecom Italia announced a €4.5m fund targeting the most innovative players in digital, mobile and green information and communications technology (ICT). The move forms part of a broader initiative, complementing its Working Capital Accelerator that has funded 19 start-ups and assigned 109 grants in Italy over a 5 year period. One of the most interesting features of the current wave of accelerator initiatives is how operator needs are shaped by local market needs and competitive dynamics. Asian operators have been bolder than many of their peers in terms of the scope of their capital arms' most recent initiatives.

A Safaricom (Kenya) fund has been launched for an initial period of two years and will offer equity and other debt solutions of USD75,000 to USD250,000 to start-ups. Eligible start-ups must have a functioning product or service, an active user base, a team in place capable of achieving goals presented, and be based in Kenya. The establishment of the fund shortly follows Safaricom's partnership with Dynamic Data Systems to launch M-Pesa payments tracking mobile application M-Ledger.

Telekom Malaysia (TM) is to launch its first startup accelerator programme in January 2015 in a move that aims to address two problems with a single solution. The country's largest telecom and ISP giant wants to find innovative solutions to problems within its own verticals and at the same time help nurture startups in Malaysia's ecosystem. TM, with its 2.3 million fixed broadband customers and 500,000 SMEs, is looking for a way to stay relevant in its market through its new Innovation Exchange, but at the same time wants to position itself as a player in Malaysia's emerging startup ecosystem that is taking on an ever-larger role in Southeast Asia.

TM is spending around US$1.8 million to US$2.3 million on creating solutions for its verticals. But the development time is slow and often the company discovers only after the spend that it doesn't require the whole suite of services developed. This is where it hopes to bring in innovative startups. One such example is Ebizu, a startup that TM is currently working closely with. It is a point-of-sale system provider that was just a small startup when TM discovered it. Now TM is packaging Ebizu's product together with its SME package known as 'Shop In A Box', which offers a range of tools for merchants. No doubt that it's important to be in contact with disruptive innovators creating products that Telcos might want to sell to its customers.

In the era of declining voice profits Telcos should be looking for innovative products to bundle together with their core offerings .Telcos need to deepen their relationship with startups so that they can have small experiments and pilots with the products and services which would add value to their business. Actually the philosophy behind an Accelerator program is predicated on savvy thinking and the inclusivity mindset. A mindset that says I cannot develop disruptive new innovations in house because I am too busy ensuring the integrity and quality of my network : so lets outsource this activity to people who enjoy this apps / platform development game and still keep some control.

A holistic innovation program can leverage accelerator programs can also be taken, from opening APIs to key platforms and revisiting employee incentives, to realigning in-house capabilities with local market needs and opportunities. Nurturing and acquiring start-up capability is more than a "nice to have" for many players; it is a mission-critical route toward new capabilities in an age where fresh approaches to service creation have never been more valued and improvements in time-to-market are vital.

In November 2015 , César Alierta, Executive Chairman of Telefonica presented in Madrid the company's new strategic plan for coming years, under the slogan "We choose it all", with the goal of transforming the company into an "Onlife Telco", or a company that promotes connections in life for people to choose a world of infinite possibilities.

The new strategic plan is founded upon six key elements, three for value proposition – outstanding connectivity, integrated offering, and differential experience- and three facilitators, which will be Big Data and Innovation, end-to-end digitalisation, and capital allocation and simplification.

Outstanding connectivity, because customers want to always be connected, anywhere and from any device with the highest maximum quality, and Telefónica can and should respond to this demand thanks to the effort made in the last few years to transform its fixed and mobile networks.

Integrated offering, made up of a wide range of services and products anticipating and adapted to the customer's needs, with personalised proposals for both individuals and companies.

And all of this with a different type of experience based on company values and a public stance that defends the customer's interests. "Now we are going to go one step further, placing our defence of the customer's interests at the centre of our stance, which will focus on three concepts: Digital Trust, Open Internet, and Digital Access", Alierta stated.

To achieve these three objectives, the company has three facilitators: Big Data and innovation, which enable speed and development of a different type of knowledge of the customer and context in order to adapt and personalise our services to maximise customer value. The second is end-to-end digitalisation, in other words, being a 100% digital company, within and without, in order to offer customers a different digital experience, and third, capital allocation and simplification, which means allocating resources to continue advancing transformation, financing growth, eliminating complexities, and fulfilling promised goals.

"Digitalisation and Big Data will change everything, will transform all production models. They are the key to innovation, to offering our customers greater value, and to making better business decisions", explained the chairman. This exponential data growth is connected to another of the keys in Telefónica's strategy for coming years: "Customers have to recover their digital sovereignty, to own their digital footprint and consciously decide how they want to make use of their data. Telefónica will always guarantee customer privacy and will do so securely and transparently. And it shall do all of this based on the principles of responsible and sustainable business."

---♠---

Telco Challenge # 3 : The Quest for Business Transformation

Telco professionals have already realised that unless they shift to new business models based on a refined customer understanding of customer expectations they will lose even more revenue to OTT players.

The transformation required by Telcos to capture new business opportunities necessitates building on existing capabilities and carefully stitching together strategies that unlock value in the convergence battlefield.

To become fit for the future, Telcos should combine growth and efficiency and effectiveness strategically by implementing a revolutionary lean and agile modus operandi. They will need to create and participate in new business and partnership models that will transform the global industry.

Challenged with the need to transform its back office processes and networks across 65 countries, Vodafone created a business transformation pathway so effective and a business model so future-proof, it won the gold in the 2010 SAP EMEA Quality Awards.

From the start of the project, including the all-important planning and design phase and throughout its evolution, the vision and focus was on the full introduction of a single Vodafone Future State Operating Model (FSOM). Vodafone needed to enable a true FSOM to deliver a globally unified set of standardized business processes across its many networks. To make this vision a reality, the telecommunications leader created a new core business model and a new procurement center supported by a shared service organization – all underpinned by a global ERP platform.

Another example of transformation in action at Vodafone is the correlation between value management and Vodafone's desire to realize maximum potential value from its operations. Applying the value management discipline to the transformation from the beginning gave Vodafone the tools, techniques, and frameworks to achieve over US$750 million in annual cash savings by 2011. Every year since Vodafone has realized these savings, in part through shared service center standardized transaction processing and lower total cost of ownership from IT.

Most Telcos have recognised that Mobile Broadband is the biggest growth segment in the telecommunications market in Africa since the shores of the continent are awash with submarine cable bandwidth. The next generation of customers expects their services highly personalized, with names and images of their own choosing, integrated with their community and able to support self-expression and viral models

Telcos confront rapidly evolving new technologies even as they grapple with network privatization, liberalization, and significant changes in the regulatory framework. While most operators have mastered their own profitability economics and subscriber value, many lack insight on the economics of adjacent or competing business models. Consequently they fail to understand how to monetise the Data Tsunami.

The current "smokestack" view of network and service management is not well-suited to the demands of today's converged networks and services. Increasingly revenues are being lost to poor systems integration and operational problems, such as database inefficiencies. These problems are contributing to increased operational costs and delays in service introduction.

Today, we see many leading operators around the world making public statements about their commitment to transform but addressing different drivers: transforming the legacy network to support , growth assets, transforming cost structure (CapEx, but mainly Opex streamlining business processes, customer experience product portfolio, focusing on investments for growth.

Some carriers are replacing large parts of their infrastructure to remain the low-cost provider. The carriers are also changing how they go about network renewal, shifting from a product to a program focus, and creating a quite different kind of platform. (BT has been a very public example with 21C). Starting from the center and working outwards towards customers, the key items marked for renewal are:

- OpticalTransport
- MPLS Core
- Softswitches and Media Gateways
- Multi-service Edge
- 4G Wireless Access
- Wired Access va Fibreoptic
- OSS / BSS

Transformation means transforming from a bureaucratic to a customer-focused and market driven corporation capable of competing in a liberalized market .To create workable blueprints for guiding this transformation, we must take into account the web of inter-relationships between markets, products and services, core processes, technology and systems, organization, staffing, metrics, and services models. Telcos need their business strategies revised, increasing the emphasis placed on:

• Improving network performance and customer satisfaction
• Reducing OPEX associated with network provisioning and assurance (service creation and service delivery costs)
• Shortening time-to-market cycles

Network transformation encompasses strategic technology, architecture, and implementation plans to gracefully transition access, aggregation, routing, and transport networks to an all IP infrastructure;

The transformation program must set a clear vision for the entire organization as it improves operational excellence and moves towards the defined target architecture. Create a high-level transformation checklist that covers a number of tasks starting with problem acknowledgment, followed by executive buy-in and budget approval. The Transformation agenda is based on the following pillars :

1. Analysing the current and future technology investments from business and technical viewpoints. In addition review the key customer drivers and applications that generate fast ROI based on understanding the needs of target markets. Conduct a network inventory to identify and retire and redundant network elements

2. Investigate how to incorporate Web 2 / Telco 2 paradigms into the creation of your product portfolio. Web2.0 is the new generation of web services, characterized as more open, flexible and participatory in terms of creating content, applications and collaborative alliances.

3. Strategise capabilities to overcome OTT net players to make money from higher value added services by implementing " smart pipe "design. We must assess every aspect of the network from its underlying hardware and systems to its configuration, capacity, traffic flow, and survivability.

4. Perform a thorough evaluation of the operations, identify the gaps between the current methods and the future vision, define new job functions and processes, prepare a road map for transformation, and facilitate the migration process.

5. Streamline the architecture with the judicious use of web services and services-oriented architecture (SOA). In addition to streamlining the network management systems environment, platforms and tools that enable service and customer management need to be introduced.

6. People and human resource skills must be upgraded to meet the needs of the new organization structure and new IP based technologies .Nimble, efficient operations rely on modern business processes, management practices, and human resources

7. Have a clear view on the risks and formulating mitigation strategies. Risk assessment must extend beyond the usual financial and regulatory risks to consider the wider environment in which the organization operates and the full extent of its operations, now and into the future. A failure to shift the business model from minutes to bytes or misunderstanding the changing customer mindset and insufficient insight into latent data assets are such risks.

To ensure a controlled business transformation, the transformation elements need to be planned and executed holistically using Key Business Objectives as a continuous guide. The Balanced Scorecard has three major categories of metrics which can be applied to Telcos:

• Revenue and margin: providing a view of fiscal performance
• Customer experience: providing a view of the measures that impact the end-customer's reaction to the service offering, and thus drive loyalty
• Operational efficiency: providing a view of cost and expense drivers

Smaller Telcos are constantly seeking investment capital to expand their networks for greater capacity , throughput and coverage. Some of the issues to be resolved before the investors put up cash relate to supply chain economics , the effect of nex gen telco services on financial performance , 4G monetization strategies , Corporate Governance and the calibre of the management team.Spectrum assets alone does not cut it anymore. They must transform and streamline their operations in order to attract investment capital.

In MEA , fixed and mobile Telcos have not fully realized the cost-reduction potential provided by lean tools and techniques, which not only can generate savings of from 10 to 15 percent on the addressable cost base, but also simultaneously improve overall operational quality levels.This process should start with a diagnostic phase that covers network planning and implementation, operations, and management infrastructure. There are also a broad range of new OPEX saving possibilities which can be leveraged through a new generation of technologies, e. g. in the area of software defined radio networks (SDR) and self organized networks (SON).

Several Telcos have gone public with energy efficiency,power reduction, and carbon footprint reduction objectives. Verizon has established an objective for its vendors to achieve 20 percent greater efficiency by January 2009, as compared to today's equipment. France Telecom is planning to reduce the greenhouse emissions per customer by 20 percent between 2006 and 2020 and British Telecom claims to have reduced its carbon footprint by 60 percent since 1996, and has an objective to reach 80 percent by 2016. Fixed and mobile operators can foster green networks by improving network energy and cooling infrastructure, and by installing energy-saving network equipment.In terms of energy operating costs, operators must initiate active programs to identify sites with higher-than-normal power consumption and adopt specific measures to reduce it. This can include adjusting air conditioning settings, making productivity upgrades to batteries and A/C systems, and adopting low-energy designs. Companies must also investigate the transfer of expensive third-party energy contracts to players that offer better terms and conditions.

The Gurus at McKinsey believe that even that fixed-line infrastructure players will outsource network infrastructure and operation to contractors in order to optimize operating and capital expenditures (opex and capex). Making this outsourcing a success requires companies to explicitly split roles and responsibilities with the chosen contractors, establish clear reporting and interface models, and prepare, negotiate, and execute specific contracts and service level agreements.

Telecoms players can employ proprietary analyses and techniques to improve the amount of value their products deliver to customers, while at the same time, creating cost-efficient designs and calculating target costs.While personnel wages and benefits represent a major network operating cost, other high potential areas for cost cutting include site rental and energy costs. As a consequence, some operators are aggressively pursuing the renegotiation of rental contracts with an eye toward moving or eliminating those sites with the most expensive rental contracts. Considering network optimization, some operators are exploring base transceiver station (BTS) "hotels." These BTS hotels group the electronics from a number of base stations for antennae up to 20 km away.

According to the luminaries at Arthur D Little ,full cost transparency must be established in order to identify, prioritize and optimize saving measures. For this reason the first phase of the cost reduction project needs to focus on establishing a stringent OPEX/CAPEX analysis. Operational saving measures are considered as activities which render saving benefits of typically in the range of 10% p.a. These measures include obvious examples like the change of maintenance service level and backhauling optimization or reduction of product portfolio. Less obvious cost saving measures include the introduction of QoS concepts for optimized bandwidth management, reduction of room temperature in local exchanges or the ceasing of 3rd party hardware maintenance for stable legacy infrastructure where better maintenance know-how is often already available in-house.

The network operations centers of many telcos face a variety of challenges, including having to deal with technology silos, unclear ownership of network issues, lack of institutional memory that forces teams to "reinvent the wheel" time and again, and others. Given the breadth of opportunities available, operators can often capture reductions of 15 to 35 percent in NOC-related costs. Potential actions include developing a clean-sheet NOC redesign, integrating NOC services on an end-to-end basis, and instilling a problem-solving, high performance mindset within the center. By introducing optimized governance models, best-practice vendor relationship management techniques, and better negotiation and deal strategies, operators that revisit mobile outsourcing typically identify the potential for an additional 5 to 10 percent in cost reduction, representing 2 to 3 percent of total costs.

Mobile operators can make use of the rich variety of customer data they have on hand to improve their network quality and target investments on a site-by-site basis. Taking this type of highly granular review of network performance metrics, site utilization, and commercial performance will enable leaders to pinpoint spending requirements.By using techniques such as network caching and CDN, operators can reduce one-on-one network downloading and hence, network load and form partnerships with broadcasters to share investments and build a large-scale, secure, single network infrastructure.

Revisiting the organization's zero-based budgeting decisions using the latest insights and business priorities can reveal new opportunities to reduce investments andcosts in areas where an operator's market share is below critical thresholds. Competitive pressure is eating into revenues and causing a spike in subscriber acquisition and retention costs (SARC). The e Channel is an opportunity in cost reduction since it enables signing up higher ARPU subscribers at lower cost of sales and cost .Key success factors for an effective e channel strategy include : create a compelling channel experience (exclusive offers) ; build a solid IT infrastructure (SOA) and revamp the organizational structure (self-contained empowered eChannel teams).

Customer self-care has been shown to reduce costs in customer contact centres by as much as 20%. This sector has evolved to become a full set of self-service capabilities that includes customers researching and buying through self-directed channels. Buying via these means has been shown to increase the revenue per user by as much as 18%, when the Telco provides an effective self-service interface for the customer.

A basic cost reduction mechanism and culture across all staff must be in place (e. g. personal target setting, cost transparency, etc.). The challenge is to embed an organisational discipline that will constantly challenge the existing cost basis. The benefits of creating a performance-driven culture within Telcos come from its capacity to amplify subsequent improvement initiatives – in effect, supercharging them. However, as with most transformational approaches, "getting there" requires strong, visible commitment from company leaders, solid organizational planning and training, and communication clarity.

One thing for sure : solving the cost/efficiency puzzle requires a wholistic mult facet approach that will target the right levers to optimise cost even as capex is injected into building high speed IP based broadband networks.

Over the years Telco IT infrastructures have evolved into expensive, complex collections of monolithic applications interconnected with specially built point-to-point interfaces. Transforming OSS/BSS platforms and partially rebuilding with more modern and harmonized platform components leads to considerable savings in the long run: reductions in time, effort and costs spent on system integration, administration, maintenance and training.

BSS systems (typically including billing and CRM), have always been separate from OSS systems (such as resource management, service activation, provisioning, fault management, etc.), which included having separate business processes and people. For example, revenue focused BSS was always run by the IT department, and cost-focused OSS was run by network operations. This traditional binary approach would have likely continued to be sufficient if not for the major transformation the telecommunications industry is undergoing, where service providers are becoming retailers of multimedia and entertainment services.

An enduring vision of harmonized OSS architecture is inspired by the TMF Lean Operator Initiative and based on four key areas:

• CSP's process architecture: The CSPs' business processes can be supported by introducing modifiable operator process templates out-of-the-box and enabling a higher level of automation in their daily routines.

• Common information architecture: stepping away from "stove piped" data and supporting shared information and data models. This enables OSS/BSS level application interoperability through Common Information Models.

• Modular application architecture: will bridge the gap between service and resource management applications. A high level of modularity allows flexible solution building: it enables easier maintenance, allows changes on one component without affecting others and allows new components to be added as required.

• Application integration architecture: Interoperability and time to market is improved through compatible interfaces, common information models and through leveraging partner ecosystems and productized adaptation libraries.

Bear in mind that the transformation to the next generation OSS is a revolution, nor is it fixed to a particular date or year. As current OSS systems are crucial for the operation of a network they cannot be replaced overnight. The transformation and migration will need to happen gradually, making the challenge even greater – old systems cannot be turned off before new systems are in place. To mitigate these risks, future needs must be anticipated in advance and OSS architecture must be designed to fit with future requirements from the start. OPEX for the legacy OSS needs to be reduced to make room for new investments and replacement of the old functionality. OPEX reduction takes many forms, including:

• Removal of old, redundant OSS applications and systems
• Streamlining of functionality in legacy OSS
• Replacement of bespoke/ customized systems integration work with standards-based software and off-the shelf mediations
• Selective freezing of legacy OSS applications and systems
• Encapsulation of functionality and making it "service aware" with SOA
• Effective use of key OSS systems, moving functionality to these and taking other systems off-line

The Enterprise Architecture Model describes the elements of business – strategy, business cases, business models, processes, supporting technologies, policies, and infrastructures that make up an enterprise. It also provides means for governing the enterprise and its information systems, and planning changes to improve the integrity and flexibility. In other words, Enterprise Architecture crystallizes the organization – what it has to do and how – to be as efficient and productive as possible.

In the Enterprise Architecture, the business quadrant handles the value chain aspects relevant to the business as a whole: where to improve the business efficiency and develop new value propositions and how to increase efficiency and competitiveness of the business in the context of its environment: markets, competitors, legislative and environmental aspects, influences and impacts.

With Enterprise Architecture (EA), new opportunities and capabilities will raise some real competitive advantages for Telco operators. Architecting the future state of EA is the heart of the entire process. The goal is to translate business strategy into a set of prescriptive guidance to be used by the organization (business and IT) in projects that implement change. As such EA is a process discipline. Done well, it becomes an institutionalized part of how an organization makes decisions to direct its investments, such that the chosen business strategy will be realized. Usually a system is seen as a necessary cost to make the business – not anymore and certainly not with EA !!

Managing IT complexity to support business strategy is a big challenge for enterprise architects at large companies when a company has global operations, as is the case for Telstra, an Asia-based telecommunications firm. However Telstra's enterprise architecture (EA) team addressed its challenges by focusing on customer engagement, improved agility, and global business strategy enablement.The EA process bridges the gap that otherwise exists between business strategy and technology implementation. High-performing organizations are process-disciplined which is lacking in many of the Tier 2 Telco operators. In turn, every high-performing process must be defined and documented, have process owners and be closed-loop with governance in place.

Activities in this phase of the architecture process include but not limited to :
• Scoping the EA program and the next iteration thereof in terms of breadth and depth, which is known as defining what is meant by "enterprise"
• Gaining executive sponsorship and support
• Conducting stakeholder analysis
• Identifying the EA leader or chief architect
• Building and chartering the "EA team," which will own and facilitate the EA process and establishing clear roles and responsibilities
• Assessing organizational readiness and EA maturity
• Developing an initial communications plan, communicating the role of EA and setting expectations of individuals participating in the process
• Establishing a plan for setting up a governance mechanism
• Defining measures of success to articulate value delivered

Currently many Telcos are burdened by a wide range of systems 'isolated' for the operation of its business. This reality does not allow effective sharing of information between systems and / or applications. In recent years they have acquired several technologies were acquired from different manufacturers and suppliers, most could be considered islands of information and technologies. Today's service providers must close the gap between their Customer-facing BSS and network-facing OSS. With Enterprise Architecture (EA), new opportunities and capabilities will raise some real competitive advantages. As an example, let's consider a set of typical (separated) systems:

1. Automation system: The principle behind this is to improve efficiency, automating several steps (or all steps) of certain tasks. Since the tasks are automatic, the delay is caused by latency of the system itself, leading to execution of thousands of tasks per second instead of seconds/minutes spent in each task.

2. Customer segmentation: Groups people according to attributes that store information relevant for understanding customer behavior, and can be used to predict the probability of acceptance or refusal of a certain product or probable churn.

3. CRM tool: Contains all customer information and supports the call center team in customer interactions.

The new reality in the Telco industry is that the basic currency of the smart network is DATA. The move to all-IP networks and the technology that has become available means that operators can collect more data than ever before from all points between their core networks and their end users and exploit it in ways not previously imaginable. Excellence in IT architecture is fundamental to efficiency and effectiveness, touching every aspect of a telco's business performance. Fortunately most senior

Telco execs have already realised that the integration of information systems, collecting, consolidating and making available all data efficiently is an essential requirement to ensure the viability and competitiveness, avoid errors and waste, improve efficiency and increase the success factors internal. As such any strategic plan to transform the BSS/OSS using EA must :

• Align the needs of information systems with business strategy,
• Monitor the rapid evolution of Information Systems,
• Rationalize and monetize investments in Information Systems
• Prioritize solutions to develop in the future according to the business strategy defined by the company
• Controlling the proliferation of systems / applications isolated and walk to the integration and overall management of Information Systems

The IT industry has embraced the concept of a Service-Oriented Architecture (SOA) as a standardized, more efficient way to build enterprise IT infrastructures. I believe that SOA, together with a revised enterprise business process, is the right way to build BSS and OSS applications, because it supports more agile internal operations, enables interoperability among new applications, and can be used to leverage existing BSS and OSS assets by adapting them to the SOA model. However to yield genuine value, an architecture transformation also requires a substantial shift in mindset.

Mckinsey are right about the fact that transforming a large telco's enterprise architecture management function to deliver maximum value is a Herculean task and multi-year effort requiring full buy-in from the business side. There is no uniform panacea for success. But the impact on costs and business performance can be huge once the enterprise architecture moves toward a uniform blueprint with consistent management across domains. Tariff changes take days rather than months. Customers can be tracked across their lifecycle and targeted with optimally customized offers, while network utilization soars.

Today, communications service providers deliver traditional and IP services that span voice, data, video, content, prepaid and postpaid, fixed and mobile. To succeed, service providers must change to an infrastructure that supports new business models, real-time customer interactions, and new partners and channels. Driving this transformation are underlying business applications such as billing and revenue management (BRM), customer relationship management (CRM), and enterprise resource planning (ERP).

 The new enterprise opportunity is a credible source or new revenue but there is still great confusion over where to start, how to scale and which divisions within telcos should be targeting these markets. Cloud Computing has captured the interest of enterprise customers because it offers flexibility and an ability to control costs. It fits well with the new Telco philosophy as it is about supplying assets as services.

When implementing billing and customer management platform several factors come into play : integration with existing systems, managing multiple billing systems, installation and service scheduling; trouble ticketing, billing blended services, managing channel relationships, automated provisioning, customer self-service, provisioning flexibility, and reporting. In fact, when properly integrated, a service based billing/CRM/Provisioning engine can evolve from an expense item to a revenue engine. If properly managed, efficient billing and CRM can translate into lowered costs, increased cash flow and increased profits.

Telcos are under constant pressure to optimize operational costs, gain agility and offer superior services to customers. In a wicked competitive environment , containing costs, streamlining operations, retaining customer loyalty, and maximizing the Average Margin Per User (AMPU) becomes a business imperative. Short product life cycles and over¬heated marketing are overwhelming the operators, which resort to ad hoc solutions that appear to offer custom¬ers what they want, but in fact mask additional costs. These costs, however, may not show up until further down the service delivery chain in other areas of the business, where their root cause may be understood but cannot be addressed across functional boundaries.

Complexity is a fact of life for telecom operators, but it is also a cost driver. Legacy systems are maintained alongside next-generation networks .The complexity that has overtaken the telecom business has resulted in organizations with technology frameworks, tariff structures, and product catalogues that if plotted on a chart would resemble a Jackson Pollock painting.One European oper¬ator found that it was offering

20,000 different tariffs to 15 million custom¬ers in one country; after it streamlined its processes to respond to real customer needs, the number of tariffs was reduced to 8,000 !! Analysis by the BCG shows that Telecomms is one of the most inefficient industries with over 40 % of its cost base gobbled up by waste in various telecoms processes.

So what do we do ?? To start with look at TM Forums's Business Process Framework (eTOM) : a widely deployed and accepted model and framework for business processes in the Information, Communications, and Entertainment industries. As a key part of TM Forum's Frameworx, the Business Process Framework represents the whole of a Service Provider's enterprise environment in a hierarchy of process elements that capture process detail at various levels.The Business Process Framework (eTOM) describes and analyzes different levels of enterprise processes according to their significance and priority for the business . For CSP's , the Business Process Framework serves as the blueprint for process direction. Here are some case studies to vindicate e TOM's effectiveness.

Qwest wanted to transform its service delivery to shorten the time-to-market for new products, including cloud services, reduce its operating costs, and have visibility and traceability from products to services to resources. It was also determined to reduce individual service component redundancy and enforce Qwest's high standards for the overall customer experience.To reduce investment risk and prove the viability of what it wanted to achieve, the operator and its partners turned to TM Forum's Frameworx and Catalyst Program before it embarked on the transformation. Within a year of the deployment Qwest saw a 4 percent increase in revenue, a 5 percent cost reduction, a 25 percent improvement in new product deployment cycle times, and a decrease in unique provisioning and assurance job steps.

Magyar Telekom's project to convert a legacy provisioning system into a single platform successfully enables the provisioning and activation of multiple product lines. The implementation relied heavily on TM Forum Frameworx and is delivering many benefits. They include cutting service activation by 20 percent and increasing the ratio of successful automated activations by 30 percent. Time-to-market for services was reduced by up to 20 percent, while the time needed to integrate new network management systems fell by 30 percent. When manual interventions are needed, they take 70 percent less time. The deployment of a zero-touch home gateway has lessened field force activity by 30 percent. New and existing services are being migrated to a new platform, and CRM will be enhanced to support trouble ticketing and the management of service level agreements.

Concurrently with e TOM an approach called Lean Six Sigma (used so succefully in corporates from the Fortune 500 financial , manufacturing and service industries) can be implemented to cut waste and inefficiency in Telco processes. Lean Six Sigma is a managerial concept combining Lean and Six Sigma that results in the elimination of the eight kinds of wastes / muda (classified as Defects, Overproduction, Waiting, Non-Utilized Talent, Transportation, Inventory, Motion, Extra-Processing) and provision of goods and service at a rate of 3.4 defects per million opportunities (DPMO). Lean Six

Sigma utilises the DMAIC phases similar to that of Six Sigma. DMAIC (an abbreviation for Define, Measure, Analyze, Improve and Control) refers to a data-driven improvement cycle used for improving, optimizing and stabilizing business processes and designs.

There are 4 overarching strategies in the endeavour to create a LEAN MEAN Organisation.The categories can be seen as structural, transformational, changes with high complexity. Pursuing any of these should not be seen as a replacement to the first strategy of continuous improvement – there is always something more that can be done to improve the efficiency within the business as it is today.

1. Improve cost efficiency and productivity through automation, centralisation, market differentiation and reengineering of work processes (including partnering)

2. Realise national economy of scale by mergers & acquisitions with competing operators (including network sharing)

3. Achieve international economy of scale by implementation of cross-border working processes

4. Leverage national economy of scope by integrating fixed, broadband, TV or mobile businesses

Today, every aspect of your telco's operations needs to be measured against the touchstone of COST EFFICIENCY to ensure it brings profits. Whether it is investment in Transformative IP programs, in merging and acquiring enterprises, or in outpacing competition; eliminating redundancies and optimizing processes is essential. Getting rid of people (to cut costs) hardly requires imagination unless senior execs have overloaded the Company with relatives , buddies and PA's who double up as girlfriends !! Using both e TOM and Lean Six Sigma Telcos can cut costs INTELLIGENTLY in a variety of areas as identified by the gurus at AT Kearney :Network, marketing, and IT : These three areas have the most potential for optimizing operational and capital expenditures, typically by reducing complexity.

Supply chain and procurement : Some Global Telcos aspire rapid international growth—often through acquisitions presents plenty of opportunities to improve supply chain and procurement capabilities. By standardizing purchasing requirements and internal technical specifications, consolidating volumes, and optimizing deals with suppliers, operators can cut costs without affecting core operations.

1. Back office : Consolidating back-office functions such as HR and finance, potentially by establishing central or regional shared services, can increase efficiency

2. Information technology : Centralizing IT services and standardizing or consolidating applications and hardware can substantially reduce costs and often improve service.

3. Infrastructure sharing : Sharing infrastructure among operators is another way to optimize costs and leverage economies of scale. For example, Bharti, Millicom, and Vodafone (Spain, Germany, U.K., India, and Ireland) have shared networks with other operators. In Sweden, 3 and Telenor's joint venture, 3GIS, covers around 70 percent of its network with shared infrastructure.

Outsourcing: Outsourcing non-core activities, such as fleet services and facility management, can improve efficiency and allow more management focus on customers. Newer outsourcing models include managed capacity, where an outsourcer is paid on a variable utilization or capacity basis. These models, besides increasing efficiency, reduce risk, and limit financing needs while fundamentally shifting the focus from operations to customer experience and partnership management.

Energy efficiency : Energy efficiency can cut costs while reducing environmental impact. France Telecom-Orange, for example, is aiming to reduce energy consumption by 15 percent between 2006 and 2020. By the end of 2010, the group had fitted more than 8,000 network sites with optimized ventilation systems, cut energy consumption at data centers, and installed solar-powered base stations (mainly in Africa and the Middle East).

Telenor, for example, reduced its software licensing costs by 34 percent by replacing local licensing agreements with global deals.Telcos will need to use the full scale of their groups to create synergies, reduce external spending, and benefit from solid supplier relationships, which can bring earlier access to new handsets and network equipment.Bharti Airtel's so-called "Minutes Factory" has enabled it to target millions of pre-paid customers that would have been too costly to serve using the conventional subscriber-led model. The factory's key elements include outsourced network equipment, which enables fixed costs to convert to variable costs. Bharti's partnerships enable it to add network and IT capacity quickly and efficiently, as needed.

We know that taming complexity and streamlining operations can reduce operational costs by a third and provide customers with better service. The up-front savings achievable in the short term—six to 12 months—will cover the costs of the initial assessment that identifies how and where to implement a Lean transformation using eTOM and Six Sigma methodologies.The good news for telecom companies faced with stalled revenue growth is that there are ways to significantly reduce expenditures. Telecom carriers will have to lower their operating expenses for traditional telecom services to maximize free cash flow, which can be invested in nontraditional services. Telcos must focus on operating efficiency when offering a suite of non traditional services in the 4G data world services, as there are no "killer" applications.

At a high-level meeting during Mobile World Congress in Barcelona, senior leaders from eight major mobile operator groups, serving 551 million mobile connections across Africa and the Middle East, resolved to cooperate on network infrastructure sharing initiatives that recognise the profound impact of mobile broadband and Internet services on the citizens of both regions. The participating operators made this commitment in order to provide Internet and mobile broadband access to unserved rural communities and drive down the cost of mobile services for all sections of the population.

This initiative basically echoes the GSMA's call that telecom regulatory frameworks should encourage flexible commercial sharing arrangements and facilitate access to government-owned assets at preferential rates to help speed up the roll-out of new networks and support the business case to extend mobile networks into rural areas. Regulators should consider the competitive advantage that sharing of towers could provide in their respective markets. However, what they have to bear in mind is the fact that new and smaller operators will be incurring lease payments as an operating expense with relative lower risk, whilst the large and incumbent operators are still recovering the capital expense incurred in erecting the towers.

So what is really driving this network sharing phenomena apart from the altruistic motives of bridging the digital divide ?? Well how about the fact that increasing competition, along with investments in ever-changing technology, which has been pushing telecom operators towards new ways of maintaining margins. Since building and operating infrastructure is a significant cost for operators ,network sharing it is the ideal way to roll out infrastructure quickly and efficiently in low ARPU rural environments. operators can rely on a single set of infrastructure for their network. According to experts the estimated Capex savings resulting from tower sharing in the Middle East and Africa region amount to USD 10 billion. Quantifying and realising these savings requires a rigorous business plan and a meticulous execution controlled through appropriate contract governance structures and well-defined service level agreements.

Currently the most commonly shared infrastructure among operators is passive infrastructure, as it is easier to contract its set-up and maintenance. Sharing passive infrastructure only, means that newer operators still need to set up their own transceivers and other transmission equipment.Passive infrastructure sharing (commonly referred to as tower sharing) has attracted significant interest from both operators and tower companies. Companies like Helios Africa, American Towers and Eaton Telecom are already working to gain first-mover advantage by pursuing tower acquisitions in the region. Over the last 2 years, the tower business has grown into a fully-fledged industry in Africa and the Middle East.

Passive infrastructure sharing requires the consideration of many technical, practical and logistical factors although the principle is simple in theory. Any potential impact must be assessed and fully understood before sharing commences to ensure that there are no adverse effects on the operation of the site and the supporting network equipment and systems. Operators must consider items such as load bearing capacity of towers, azimuth angle of different service providers, tilt of the antenna, height of the antenna, before executing the agreement. Although, tower sharing enables new entrants to scale-up faster, it exposes established players to the risk of market share loss. Furthermore, the challenges of monitoring network performance and quality will increase as control over network roll out and equipment maintenance decreases.

As passive infrastructure business has evolved into a separate industry around the world, many tower companies in the telecom industry face several challenges. These include:

• High capital requirement: Tower deployment is a highly capital-intensive activity. The installation of each tower requires an investment of USD 55 000 to USD 75 000. Thus, tower companies the world over end up being highly leveraged

• Regulatory clearances: The first step should be to ensure that the regulatory authority is in favour of infrastructure sharing. Projects may stall because of delays in regulatory clearances. Apart from dealing with telecom regulators, tower companies also have to deal with other governmental bodies such as municipalities, forestry departments and environmental departments.Hurdles in obtaining clearance from a multitude of governmental bodies are often cited as reasons for delays in several site installations across developing nations. Since most of them are regional in nature, tower companies have to deal with quite a few governmental offices scattered across the country

• Operational cost optimisation: Although operational costs such as power and fuel are generally passed on to the operators, these are usually subject to agreed maximum limits. Thus, tower companies must work towards building controls to limit operational costs. Tower companies also face the problem of finalising the cost-sharing percentage and building a technology road map.

• Handling of local issues: Tower deployment and operation involves dealing with location-specific issues, including dealing with the landlord and local authorities, and running operations across a variety of geographies and terrains.

According to KPMG , the accounting treatment for infrastructure arrangements would depend on the model applied and the structure of the transaction. Accounting for these arrangements could be complex and a detailed analysis of the substance of the arrangement is required. Operators could:

1. Retain the infrastructure assets on their books (typically if risks and rewards of ownership are retained)

2. Derecognise the infrastructure assets (typically if risks and rewards of ownership are transferred to the third-party tower company)

3. Recognise a portion of the assets (typically if there is joint control over the asset)

In the US , TowerStream has formed Hetnets Tower Corporation, which will offer wireless carriers and others a range of shared infrastructure services and access for mobile wireless Internet services. They believe the explosion in mobile data in urban markets is driving a migration to small cell architecture, and the major carriers are presently focused on the densification of their networks. With the rise of mobile data placing a tremendous demand on the networks of the carriers, TowerStream concluded that its Wi-Fi network can serve carriers' data offload needs. To serve this need, Hetnets Tower Corporation rents space on street level rooftops for the installation of customer-owned small cells, which includes Wi-Fi antennas, DAS, and metro and pico cells. Channels on TowerStream's Wi-Fi network are available for rent for the offloading of mobile data.

Concurrently there is a growing industry in green technology that specialises in producing energy from renewable sources or with zero or reduced carbon impact. Such technologies include solar power, wind power, wave power and bio fuels. Operators should be in a position to benefit from these technologies as the amount of power they can generate continues to improve. Vendors have already successfully trialled a combined solar and wind powered base station in various African countries, which not only reduces the environmental impact of the site but also makes it more feasible for operators to deploy sites in remote regions by negating the need for traditional power supplies or maintaining a fuel generator.

Network roaming can be considered a form of infrastructure sharing although traffic from one operator's subscriber is actually being carried and routed on another operator's network. However, there are no requirements for any common network elements for this type of sharing to occur. As long as a roaming agreement between the two operators exists then roaming can take place. For this reason operators may not classify roaming as a form of sharing as it does not require any shared investment in infrastructure. When roaming agreements come to an end they can be renegotiated either with the existing host network or another operator with minimal effort and transitional impact.

Network sharing is increasingly favoured by progressive policy makers as a way of ensuring more rapid provision of 3G services and on environmental grounds. On this basis the European Union has consistently ruled in favour of permitting passive network sharing and more recently also national roaming under the caveat that competition rules are respected. The sharing of sites and masts, national roaming and RAN sharing tend to impact coverage, quality of service and pricing of services to consumers positively, as the cost saving characteristics of infrastructure sharing allow for increased efficiency.

One of the largest commercial infra sharing deals occurred in August 2004 between Telstra and Hutchinson in Australia This was cleared by the ACCC (Regulator) who assessed the benefits outweighed the potential competitive impact. Telstra paid $450 million to Hutchison Telecommunications Ltd for a 50% share in ownership and operation of its 3G radio access network infrastructure. The cost to Telsra of building a network over four years would have been $900 million to $1.0 billion. Telstra stated the deal was undertaken to save on costs of entering the 3G market and that they scored a tried and tested network at half the cost. Surely the same logic would apply to 4G LTE networks in MEA.

How can an organisation transform to become agile, do new things, whilst continuing to do the stuff it already does? There is no one simple answer. In an ideal world one would simultaneous transform senior management's mindset, strategy, structure, systems and culture. But in reality that's hard, if not impossible, to do. It is more realistic is to seek to change areas over time, building on successful initiatives. For example, an operator could make one business unit, process or system more agile, measure the results, and use the proof of success to lead senior management to embrace the concept.

Operators that do not undergo a lean transformation, however, will find themselves unable to compete. The decision to adopt a lean and mean approach needs to be made before you become extinct !!

--♠--

Telco Challenge # 4 : The Quest for Data Monetisation

Telcos are being impelled to deploy LTE as quickly but the the the same time refresh their traditional operational support systems and business support systems while simultaneously launching innovative products that can justify their heavy investments in LTE.To realize 4G's full potential and begin seeing a return on their extensive investments, service providers need back office systems that are just as robust, fast, flexible, and smart as the network. LTE-enabled services are driving new revenues for operators but also pushing them to meet the increasing data demands of customers signing up for high speed mobile broadband offerings.

The primary facets of a 4G monetization strategy involve speed to market, real time, flexibility of offer and charging/policy convergence, customer self-service over other interaction channels and the need for real time marketing analytics. To enable richer more compelling personalised services Telcos must deploy networks with capabilities, such as mobility, messaging, location, presence, profile and call control, and combine these with internet-style services such as social networking, search, advertising, direct marketing and mapping,

"The industry focus in the coming years will be on personal data, connected living, digital commerce and networks. To realise the necessary developments in these areas, "huge investments of $1.7 trillion need to be done from now until 2020",

(Jon Fredrik Baksaas,CEO of Telenor and chairman of the GSMA)

Recent months have seen an increasing number of wholesale telecoms operators and other companies enter the market with inter-carrier IP exchange (IPX) services. Several developments will drive the long-term demand for IPX services, including the launch of 4G networks based on LTE, the growing demand among VoIP service providers for high-quality transit and termination, and the ability of IPX to support new services, including videoconferencing and HD voice.

By providing hub-based interconnection to telecoms operators, content providers and other companies, IPX service providers aim to offer a private international network for exchanging both IP and legacy traffic that is separate from the public Internet. As much as 25% of the wholesale value of traffic flowing over IPX networks is expected to be generated by new and value added services, with LTE interoperability, roaming and voice services set to be strong drivers for growth, according to telecoms analyst house, Innovation Observatory.

IP exchange (IPX) is a telecommunications interconnection model for the exchange of IP based traffic between customers of separate mobile and fixed operators as well as other types of service provider (such as ISP), via IP based Network-to-Network Interface.The intent of IPX is to provide interoperability of IP-based services between all service provider types within a commercial framework that enables all parties in the value chain to receive a commercial return. The commercial relationships are underpinned with service level agreements which guarantee performance, quality and security.

French operator group Orange's wholesale division has expanded its IPX (IP exchange) with a Diameter signalling offering allowing customers to introduce LTE roaming across Europe, the Americas and Asia.The operator said that it has an ambitious program in place to extend it's LTE Signalling capability through direct connectivity and peering agreements. According to the operator the Diameter-based LTE signalling service enables operators to provide end users with improved QoE on 4G networks while roaming outside their home country.

The Telenor Global IPX enables operators and partners to be connected through one IP connection. The IPX interconnect solution allows for an optimized, flexible, secure connection and guaranteed Quality of Service (QoS) for voice and mobile services. This is an important shift to improving Global Roaming Quality for the end customer. The Telenor Global IPX Service is provided over the IPX Compliant MPLS/IP Network and includes the basic framework designed by the GSMA which incorporates four main principles:

1. Premium quality

2. Secure environment

3. Flexible for all services

4. Cascading payments

To establish LTE data roaming worldwide, two architectural issues need to be addressed. First, it's not that easy for an LTE operator by itself to complete interconnectivity with its global roaming destinations. Instead of current GRX and bilateral TDM links, roaming between LTE/EPC networks requires an efficient international exchange mechanism to integrate all the IP-based services originated in EPC/IMS and to interconnect them among numerous LTE operators worldwide. Besides, the existing GRX defines only best-effort IP transport, a QoS guarantee scheme between visited and home networks is required to let LTE subscribers enjoy international services with the same quality as domestic services.

Telstra Global is developing its capability to become a leading IPX provider in Asia-Pacific on a base of service flexibility for LTE roaming and service-aware traffic management. Telstra Global's IPX strategy for LTE roaming borrows lessons from the cloud by offering its customers the ability to burst traffic in peak times. Telstra Global is also supporting the dynamic allocation of classes of service (CoS) to address the increased requirements of the content- and applications-based traffic expected to flow over next-generation networks. Telstra Global sees basic LTE data roaming as just the beginning of its IPX ambitions

Second, instead of the legacy SS7 MAP in GPRS roaming, roaming signaling in LTE/EPC uses Diameter on IP to exchange subscribers' authentication and location update between visited and home networks. Diameter is the Authentication, Authorization and Accounting (AAA) protocol succeeding to RADIUS. Unfortunately, interoperability of Diameter technology hasn't yet reached full maturity. Actually, different vendors' EPC nodes sometimes can't successfully communicate to each other due to different implementations of Diameter signaling despite 3GPP standards.

Spectrum fragmentation is the big elephant in LTE roaming room .3GPP identifies more than 20 LTE Frequency Division Duplex (FDD) frequency bands and more than 10 LTE Time Division Duplex (TDD) frequency bands. According to Analysts, LTE operators have committed to launch in at least 13 bands, and with as many as 10 bands being proposed in a single region. It increases the cost for LTE operators to provide global roaming coverage with multiple bands, as well as for manufacturers to develop roaming-compatible handsets.

To solve this problem, the industry has to seek a set of common bands for international roaming or has to wait for a handset that supports a sufficient amount of frequency bands. Whereas the currently popular bands (800MHz, 1.8GHz and 2.6GHz) are likely to be applied for international LTE roaming for the time being, a global consensus on common bands for international roaming is necessary for sustainable growth of LTE and its roaming.Despite the spectrum problem, a lot of MNOs, IPX providers, and vendors are preparing and testing LTE roaming now.

Therefore, international LTE roaming will certainly expand particularly in the popular frequency bands. Thus, international LTE roaming on IPX embodies interconnectivity of

IP-based telecommunication services among LTE/EPC ecosystems and reflects the beginning of new experience on global coverage of high speed mobile computing.

Recently SAP AG announced an IPX peering agreement with Etisalat UAE, the largest telecom operator in the Middle East and Africa, to deliver LTE roaming traffic to all mobile operators. This strategic LTE roaming and diameter peering agreement will help Etisalat operator companies to interconnect with SAP Mobile Services' strong IPX customer community and launch LTE roaming quickly.

Whilst LTE is spreading across the globe,(222 commercial LTE operators in 83 countries GSA Evolution to LTE report: October 17, 2013) the question remains how to best leverage the speed and capacity that LTE brings to create value for customers and generate more revenue.).The introduction of LTE should be used to enhance the value offered to consumers. Operators should introduce new pricing schemes (e.g., leverage quality of service differentiation), enter new markets (e.g., fixed broadband, services such as video streaming and calling), and gain market share (e.g.,in mobile broadband and in sub-segments such as the high-end customer segment).

Based on improved technical features and economics, operators can selectively utilize customers' greater willingness to pay to counter the price reduction trend. New mobile Internet customers need to be educated on the superior service quality on offer and drawn to the brand with introductory offers to counteract low willingness to pay.

With LTE, tiered pricing is evolving from volume-based tiers to speed-based tiers with data packages based on varying speed entitlements, but often a combination of both.Bear in mind that implementing speed-base packages requires sophisticated policy control architecture connected to an online charging system. Typically when a subscriber begins a data session it triggers a message to a Policy Manager which reviews the Subscriber Profile Repository for the customer's entitlements. The Policy Manager enforces the rules through the Enforcement Point (a network packet gateway or deep packet inspection node). Policy management can play a significant role in this initiative by providing operators with a way to offer their customers innovative, personalized bundles of services that are consistent across multiple channels. Via initiatives such as cross-sell/upsell, highly targeted services and creative packaging, operators can provide their customers and prospects with more value and, in doing so, drive loyalty behavior

There have been a lot of discussions around unlimited data offers, with many analysts arguing that they were unsustainable and that operators needed to create more differentiated and value-based offers. Swisscom has actually been using speed as the differentiating value whilst offering unlimited data on LTE – with its Infinity tariffs launched in July 2012. Based on the fact that the value of speed is easier for customers to grasp than data volumes, this approach gives customers the worry-free "unlimited data" that they like, whilst still differentiating to optimize revenue.

As a result, Swisscom has seen its overall ARPU grow, although it initially declined its market share is increasingly growing. KPN Holland are packaging speed as the main promoted value of LTE. It enables them to create differentiated packages where speed is used to create perceptible value – and encourage customers to upgrade to higher tiers and 3G customers to switch to LTE for a premium.

3 Austria follows the same strategy as Swisscom with its Hello and Hello Europe tariffs that include unlimited data with differentiated speeds and mobile TV channels. However, whilst offering unlimited data, both Swisscom and 3 Austria restrict the speed to 64Kbit/s once a certain amount of data is consumed; 3 Austria clearly highlights this to its customers and shows for each tariff the full speed data volume in addition to the maximum download and upload speeds. However, 3 Austria also offers a premium package with unlimited and unthrottled data speed.

Mobile US for instance offers high speed data volume tiers with its "Simple Choice" plans and then throttles to 2G data speed once the high speed data allowance is exceeded, until the next billing cycle. In effect, this approach is equivalent to offering unlimited data to customers at differing speeds. Orange France also throttles the speed when the data allowance is exceeded. Optus in Australia automatically moves customers to the next tier for the rest of the month. AT&T sends data usage alerts to its customers and then charges when they go over their allowance.

In terms of roaming, T-Mobile US offers unlimited data while roaming as part of its Simple Choice plans; however the speed is limited to 128Kbit/s , the aim being to upsell "speed packs" when higher speed is required. The plans are subject to fair usage rules and also include unlimited SMS while roaming. This approach is very attractive to customers whilst enabling the company to continue to drive roaming revenues by upselling speed. Temporary speed boost offers may also bring additional revenue from travelers by allowing them to quickly download movies or games to enjoy during their trips.

LTE has brought the opportunity for operators to upsell high-speed add-ons to supplement a base package or temporary speed boosts, enabling customers to purchase higher speeds as required. For example, A1 Telekom Austria upsells high speed options to its customers in addition to basic packages that incorporate lower speeds. A1Telekom Austria offers a data add-on (Eur 8.25) to its business customers, with night time access (22:00-08:00) to unlimited data, at full speed as per their base package. This is an interesting approach to managing speed and network resources whilst offering the always popular "unlimited" to create a new data revenue stream.

EE UK applies a speed cap of 30mbps to all new customers not on its premium post-pay package "4GEE Extra" while premium customers can enjoy speeds of up to 150 mbps. Their packages also include unlimited minutes and texts as well as free roaming calls and texts to selected countries and includes the Deezer music service. Telia, the world's first LTE operator offers its 4G post-paid "Mobil Komplett" customers a speed up to 100 Mbit/s whilst all its 4G pre-paid customers are capped to 20 Mbit/s. Telia's Mobil Komplett plan also comes with data volume tiers and shared data allowing customers to use up to 7 mobile/tablets on the same subscription. The plan includes unlimited calls, SMS and MMS.Orange France launched its low cost sub-brand Sosh which uses speed caps to differentiate from the main brand. Whilst Orange customers can enjoy LTE speeds up to 150 Mbit/s, Sosh offers a maximum of 42Mbit/s with its best package.

As the telecom market becomes increasingly saturated, operators are looking for ways to stand out without resorting to price wars. Core services such as voice or data are becoming commodities; in order to avoid being limited to charging commoditized prices, operators must be able to create and deliver services that offer additional value, but are limited by the rigidity of many legacy service creation and OSS/BSS environments.

A flexible policy management solution deployed either as a standalone implementation or as part of a larger OS S/BSS and service delivery transformation initiative, can enable value-added, differentiated services that can run on top of existing core services. While pricing can never be low enough from a consumer perspective, the ongoing quest for operators is to find a balance between competiveness and the ability to fund future investments.

Telcos are pouring billions of dollars into building out 4G networks. LTE will stimulate demand for video and media services, thanks to its lower latency and higher capacity access, and indeed mobile TV, video calling and video downloading show the strongest growth prospects. The point is that no single nontraditional service will be able to compensate for the erosion in traditional telecom revenue. Media/entertainment (including advertising), machine-to-machine (M2M) services, cloud computing and IT services are promising areas for generating revenue.

To monetise LTE , Telcos will need to define new business models and marketing strategies to drive acquisition and retention of subscribers, as well as adoption of high-value content and applications. Optimising mobile broadband economics is a complex challenge, and there's always a temptation to try to solve complex problems with one 'silver bullet'. Unfortunately this is impossible, as there are many different combinations of solutions will work at different times for different operators.Operator device portfolios are one of the most important factors underpinning network technology migration trends.

Smartphones offering a compelling user experience drove mass-market uptake of 3G mobile broadband subscriptions and LTE will be no different. In the LTE ecosystem, tablets seen to offer a greater prospect for revenue growth than dongles. It will be important to secure an attractive selection of smartphones and devices at affordable price points.To boost the need for LTE, operators can make partnerships to enable applications that allow customer enjoy the full experience of LTE.

Telcos need to adopt a balanced approach on how to monetize 4G networks by a thorough analysis of various technical , financial and commercial strategies : the desired outcome is to minimize the cost of access coverage while maximizing subscriber capture. A balanced strategic plan is founded upon :

1. Effectively reengineer the broadband business model so as to reduce costs , manage data traffic , and develop a more sophisticated approach for pricing broadband access

2. Unlock new revenue streams to justify the enormous network investments over time in the context of key customer drivers and applications (cloud , M2M etc) that generate fast ROI based on understanding the needs of target markets.

3. Collaborate with OTT players since LTE's all-IP architecture will create a more open environment for Over The Top (OTT) applications which threaten to further commoditize the network.

4. Leverage the OSS/BSS to weaponise the CRM , Billing , Policy Control systems in order to ensure that all the data traffic is accounted for and billed to the correct entities.

What service providers need to do is to offer packages based on the service or application used that can be provisioned dynamically ,rather than on bandwidth allocation. Policy management tools play an important role here. By being able to offer management tools , the provider will be able to offer subscribers packages such as ' YouTube subscription' , or ' online gaming subscription' , or ' regular surfing and email subscription . Policy(PCRF) is the brains of a network , especially for LTE networks that must make many more real-time decisions to maintain network performance and adapt the network to the subscriber.

In a Media Research survey, respondents pointed overwhelmingly to smart devices, video, and cloud services as the devices and services most likely to drive demand for 4G. Fully 40 percent already have partnerships with content providers to assure them higher quality of service (QoS) on their networks. Enhanced enterprise LTE solutions, such as videoconferencing on-the-go and remote access to business applications, can drive data consumption. Verizon Wireless is one of many LTE operators that offers 4G mobility applications and solutions for SMEs and enterprise customers. A survey shows that 67% of US businesses using LTE believe that it has resulted in increased productivity.

In order to offer a more competitive service than the OTT players KT is leveraging CCC for ICT business. CCC is a kind of domain-specific cloud technology, based on virtualisation. By unifying the platform for radio and several application services into CCC, KT can provide cross-layer optimised services between applications and radio. For example, they can utilise user contexts such as user ID, traffic content, QoS, location, and the radio environment to provide the most suitable service to their customers.

There are some valuable options for addressing the data issue from a technical point of view, offload perhaps the most valuable amongst them. However, these are not all the weapons in an operator's arsenal. They can also look to manage the impact of traffic on their networks and their bottom lines by looking at different business model and pricing options.

Operators can use the rich data experience of LTE to sell more data and develop new revenue streams. Video streaming providers such as Netflix alter the quality of video according to available bandwidth – so a 6-minute clip on LTE would consume 80MB compared with 27MB on 3G, thus driving usage. Operators are also bundling content with LTE or top-tier plans, enabling new revenue streams – for example, EE in the UK uses its film service (EE Film) to monetise data and receives sales commissions from video-on-demand provider FilmFlex.

On the revenue side, the bulk of revenue will be from 'downstream' subscription and pre-pay customers, and while helpful, that the near-term growth of new 'upstream' or wholesale / carrier services revenues alone would not be enough to cover the costs of capacity increases.Because LTE network latency is lower than 3G, operators can develop new revenue streams by selling bandwidth for wholesale services (such as utility and M2M services). Verizon is at the forefront of this with projects in sectors such as education.

MNOs can also experiment with bundling. Data sharing across devices is being offered, with the aim of monetising devices (such as tablets) otherwise lost to Wi-Fi. Tethering strategies are evolving, as operators try to monetise tethering by allowing it at as part of premium or top-tier plans. Fixed–mobile converged offerings are available and aimed at increasing revenue and reducing churn.There are many different possible solutions and different combinations of solutions will work at different times for different operators :

1. New air interfaces and spectrum will not be enough to on their own to cope with the continued rise in data traffic. Building more cells alone is not a solution, and it will be necessary to address costs and pricing

2. The challenge needs to be approached both from the network, through policy-based control including tiering and maybe traffic-shaping, backhaul optimisation, and offload through femto cells or WLAN, and from the business side with pricing, potential tiered offers and segmentation

3. Techniques have to be deployed to manage traffic to deliver customer experiences, particularly for cloud and TV services

4. Since no single method of addressing capacity issues provides a complete solution and therefore a combination of offload, traffic management and segmentation is recommended.

5. Mobile data optimization that includes content transformation is a crucial element in increasing the efficiency of data and video transport, by reducing the over-the-air payload on the RAN, and improving the subscriber experience with faster page loads and lower monthly data usage.

The companies that go to market with 4G services will have to be able to sustain them. The networks themselves will drive huge growth in data traffic. But changing business models also have the potential to explode transaction volumes. Not surprisingly, wholesale will be an important part of the 4G mix. The wholesale models most frequently cited are bulk access, machine-to-machine, and mobile virtual network operator.

However scalability and sustainability will also affect billing systems. CSPs will need to invest in the next generation of business-support systems (BSS) to manage customer-facing operations such as product, order, customer, and revenue management. 4G involves both capital and operating expenses, and those investments will have to be made simultaneously. CSPs will need to shift their perspective from cost to revenue management. For that, they will require more sophisticated policy and charging solutions.

The market dynamics of the Web2.0 will impact the LTE business models because it will be difficult to charge the user directly for the use of Web2.0 applications (that will run fast and smooth over LTE pipes) because Internet applications are associated with free usage. In view of this Telcos could exploit more indirect revenue sources : in a 'multi - sided' market structure the telco transactional platform can facilitate improved interactions and transactions between people and organisations : between advertisers and the end users.

Businesses can capitalise on 4G LTE for a wide set of applications, some of which are purely 'horizontal' while others are highly sector-specific, addressing needs unique to the industry. LTE's advantages are of greatest relevance to applications for personal communication and collaboration, CRM and project management. LTE will deliver improvements in the performance of many existing applications, and make feasible new applications that depend on reliable high speed or responsive data transfer.

Within the next decade, and probably by 2015, one trillion devices will be connected to the network – most not phones – moving the communications industry from quad-play or multi- play to "Tera-play". And LTE is one of the key enablers of a Tera-play world. For service providers, the benefits of Tera-play could be substantial with opportunities to drive revenue from an increasing number of high-value, multi-device, multi-service customers and infinitely larger personal and community networks.

According to the GSMA, LTE users consume on average 1.5 GB of data per month, which represents almost twice the consumption of non LTE data users, many of whom exceed their data limits .Furthermore, savvy operators have been innovating, with new offers and business models, to better capitalize on the demand, speed and capacity of LTE. All of these mean more revenue opportunities but as mobile pricing strategies evolve unlimited data plans are becoming less prevalent while tiered pricing models, based on different data entitlements, are more commonplace. These pricing models require operators to put in place usage safeguards and notifications for their customers to avoid bill shock besides offering other value added functionalities.

Many LTE operators are leveraging LTE speed to differentiate, by using tiered pricing to create perceptible value based on speed and encourage customers to upgrade to higher packages. Here are some examples showing different approaches to speed tiers: Telia, the world's first LTE operator, uses speed to enhance the value of post-paid offers and attract more post-paid customers. Telia applies a speed cap of 20 Mbit/s on its 4G pre-paid offers and provides maximum speeds up to 100Mbit/s to its 4G post-paid customers. Orange France uses speed to differentiate with its sub-brand Sosh; Orange France offers maximum LTE speeds up to 150 Mbit/s on its main contracts, whilst Sosh offers speed tiers varying from 14 Mbit/s 3G speeds to 150 Mbit/s ; Swisscom's Natel Infinity plans offer the always popular unlimited data, using speed as a differentiating value . However, the speed is reduced to 64kbit/s when a specific amount of data is consumed. They stated in their 2013 report: "Figures from recent quarters shows that customers switching to Natel infinity are generating higher revenues (ARPU)"

LTE has brought the opportunity to not only use speed as a differentiator but also to upsell high-speed. Some operators have been leveraging LTE speed to develop different upsell strategies: A1 Telekom Austria is offering high speed add-ons to supplement base tiered packages with lower speeds At 1Telekom Austria was also offering its business customers a data add-on (Eur 8.25) with night time access (22:00-08:00) to unlimited data, at full speed as per their base package. This is an interesting approach to managing speed and network resources whilst offering the always popular "unlimited" to create a new data revenue stream. Another approach adopted by Ooredoo is to offer tiered data packages and reduce the customer speed when they reach their limits before the end of the month. Customers can then purchase various data add-ons "Extra packs" to restore the speed.

According to research firm Juniper, almost 75% of roamers worldwide do not use data services. Most roamers choose to turn off mobile data or substantially reduce usage for fear of bill shock. The EU Parliament has tackled this issue by voting to abolish roaming charges altogether from 15 December 2015.The question therefore for many operators today is how to keep driving revenues from travellers.T-Mobile US have implemented an interesting approach. They offer unlimited data whilst roaming as part of their "Simple Choice Plan" ; however the speed is limited to 128Kbit/s. Whenever customers require higher speed, they can purchase high speed data roaming passes that provide total cost control; they offer 1 day/100MB, 1 week/200MB and 2 weeks/500MB passes. The plans are also subject to fair usage rules and include unlimited SMS while roaming.

This approach is very attractive for customers as they can access data abroad for free even though at low speed; yet the operator can still generate data roaming revenues by upselling high speed roaming passes.Real-time usage notification can prevent bill shock by empowering customers to take control of their data usage. Operators can configure different thresholds (e.g. 50% and 80%) to trigger notifications which can create upsell opportunities for those who have reached their data usage allowance. Bill Shock push notifications are superior to SMS and allow the use of graphically rich formats that better reflect the operators brand and product marketing requirements.

In a recent survey (Telecoms Intelligence BSS), 84% of operators revealed that they will invest in solutions for smart upsell offers, triggered by real-time context information, such as network usage, application access, location and more). This will help operators to maximize upsell opportunities and sales conversion rates as the offers are more relevant and timely.For example, a subscriber on a low speed data package, accessing Netflix, could trigger a high speed add-on offer; Another subscriber trying to access Facebook with no data package could trigger a data pass offer with unlimited access to Facebook. One operator successfully used this approach to stimulate data adoption by upselling real-time data passes (10MB for 1 day) to subscribers who had no data plan when they tried to consume any data. They used data passes to provide total cost control and remove any fear of bill shock .

The build out of today's 4G networks such as LTE requires a dramatic increase in computational resources to adequately deliver flexible telecommunications services to mobile subscribers. Yet business conditions also necessitate that new markets are approached incrementally. The challenge for telecom carriers is to reduce the cost of serving the first subscriber in small or cost-sensitive markets. The primary challenge in serving small LTE subscriber bases is that traditional core network architectures require high capital expenditures just to serve the first subscriber.

Networks, whether entry-level or full-scale, are traditionally built using separate network elements for each of several different functions. And most network elements have been deployed with a pair of carrier-grade servers to achieve redundancy with an active and a standby configuration. Thus, a new network with 10 network elements requires 20 servers just to provide service to the first subscriber.

Furthermore, because the network is designed to eventually support a large population of subscribers, the servers would remain underutilized until the subscriber base grows to the expected population. The ROI for small and emerging markets has therefore been limited by these high capital outlays. High operating costs for maintaining the servers and providing data center floor space, power, and cooling have also hindered new service opportunities.

The greatest opportunity for revenue growth for wireless broadband presents itself in the form of smaller markets with less than 50,000 subscribers, thereby lowering the cost dramatically to serve the first subscriber and the breakeven point in the Operator's business case . By dramatically lowering the cost to serve the first subscriber, new networks can be built on a campus or targeted community basis with new services tailored to the specific needs of these smaller, targeted markets.

According to Warren Buffet "The single most important decision in evaluating a business is pricing power..If you've got the power to raise prices without losing business to a competitor, you've got a very good business. And if you have to have a prayer session before raising the price by 10 percent, then you've got a terrible business." If this is true then it would make Telcos in the 4G world a lousy business since Telco Execs in Africa (especially) have a prayer session on how to cut prices by + 25 % every 3 months and crow about it in the Media like pontificating politicians seeking votes during a presedential election.

The Boston Consulting Group Report (The Internet Economy in the G-20: The $4.2 Trillion Opportunity,) surveyed about 1,000 people in each of several G-20 nations on what "lifestyle habit" they would give up instead of the Internet for a year, including sex, alcohol, showers and cars. Most of the results for items like coffee, chocolate and fast food were steady with averages of 70-80 percent. Japan topped the list of citizens who would make the sacrifice, with 56 percent who would abstain from sex. Brazilians were the least likely to give up sex for the web access – only 12 percent surveyed would give it up. American and South Africans were most attached to their vehicles – only 10 percent each were willing to give up their cars for the Internet. Another interesting finding was the perceived value of the Internet versus its actual cost. For instance, Americans value the Internet at $3,000. According to BCG, it's value is actually $472 – an incredible markup in price based on perception.

So in light of the above : is there any hope for Telco execs who are intent on destroying the profitability of the Industry by engaging in vicious price battles that threatens the sustainability of the Industry ???? Thats where Yield Management can provide some insights on how best to price data. Robert Crandall, former Chairman and CEO of American Airlines, gave Yield Management its name and has called it "the single most important technical development in transportation management since we entered deregulation." Ditto for Telco De Regulation.

Believe it or not the telco and airline industries have much in common in terms of perishable inventory , large sunk low marginal costs and varying predictable demand volume. Yield management principles would indicate that, as the costs of using network capacity are minimal, it makes sense to use as much of it as possible when it is available and maximize revenue from it when it is in shorter supply. Unused bandwidth is lost forever.

Yield management is the process of understanding, anticipating and influencing consumer behavior in order to maximize yield or profits from a fixed, perishable resource such as airline seats or hotel room reservations or advertising inventory or Internet bandwidth . Its core concept is to provide the right service to the right customer at the right time for the right price.This process can result in price discrimination, where a firm charges customers consuming otherwise identical goods or services a different price for doing so.It is surprising if not shocking that Operators are utilizing on average only 35 to 40 percent of their network capacity. It takes some creativity to turn this huge dormant asset into profits.There are several scenarios in which yield management can be used to increase revenues.

1. Improving Market Segmentation and Pricing : Market segmentation enables enterprises to cater products and services, including pricing, targeted at the buyers in each segment. The basic idea, which makes segmentation an effective tool to increase an enterprise's bottom line, is to charge more for products targeted at customer segments with a higher willingness to pay.

2. Monetizing Unused Bandwidth : Telcos stand to increase profits through the creative monetization of their unused network bandwidth. Similar to the empty airline seat, network bandwidth is a perishable resource with a very low marginal cost, so filling the underutilized bandwidth with revenue-producing network traffic will have a direct, positive impact on the bottom line.

3. Dynamic Pricing : Market segmentation and associated segment pricing aims to maximize profits by fixing prices at levels that are optimal for targeted segments.

Dynamic pricing encompasses adjusting the prices to changing market conditions and/or the status of the Telco's resources.

4. Increasing Profits at Peak Utilization : Although dynamic price adjustments can be used as an effective mechanism to off-load the network during peak usage, periods of peak utilization by definition are synonymous with periods of peak market demand and, as such, should be assessed for opportunities to increase revenues.

5. Reservations : The mobile network's bandwidth management functions enable Telcos to reserve bandwidth for specific customers in advance of their actual bandwidth usage. When a customer's reservation is in effect, the network ensures that the customer receives the requested bandwidth even if there are other customers competing for the same bandwidth.

6. Pricing flexibility : Real-time charging functionality provides Telcos with pricing flexibility at least on par with, if not better than that used by the airline industry. Telcos can charge based on the attributes of provided services, customer characteristics, context, network state, historical usage, etc.

Besides increasing revenue Yield management can also reduce the need to increase capacity, resulting in savings in investment for providers, which can be passed on to consumers as lower costs. The value of yield management for mobile broadband is an opportunity for service providers to manage the quality of a user's experience while achieving increased revenues in the context of the exponential CAPEX costs associated with servicing the global demand for mobile broadband services.

Across mobile networks deploying LTE radio access technology, Voice over Long Term Evolution (VoLTE) and Video calls over LTE which utilise IMS technology are recognised as the industry-agreed progression of voice services. VoLTE facilitates far richer, multi-media voice services, increasing the service quality (by offering HD Voice) and interest delivered to consumers.Concurrently with the increasing prevalence of Wi-Fi access networks and the change in attitudes relating to Wi-Fi as complementary RAT and the proliferation of Wi-Fi Calling enabled devices, the time to deploy feature rich alternative voice services has never been more favorable for Telcos.

According to the GSMA, in July 2015 422 operators deployed LTE networks in 143 countries. As the roll-out momentum continues VoLTE, which utilizes IP Multimedia Subsystem (IMS) technology, is recognized as the industry standard agreed progression of delivering voice services in a packet switch only network.The size of the VoLTE market will be over $30 billion through 2019, says ABI Research. Visiongain expects there to be 101.7 million active VoLTE subscriptions worldwide by the end of 2015. Apple's embrace of VoLTE with the iPhone 6 raises the bar and points the way for all device manufacturers to follow. ABI Research believes that VoLTE subscriptions will significantly increase year-on-year as subscribers appreciate the voice quality and demand the additional services of the all-IP 4G LTE.

So why the hoopla over VoLTE and WiFi calling ? For starters a VoLTE network has up to 3 times more voice and data capacity than 3G UMTS — and up to 6 times more than 2G GSM — and that extra capacity helps you sell more data services. To sweeten the deal VoLTE's packet headers are smaller than unoptimized VoIP/LTE, so that more bandwidth is available for data services. For instance, in a cell with 200 VoLTE calls, 4.4 Mbps is freed up for data services. VoLTE also reduces the consumption of each cell's control channels, so the cell can serve more users and increase throughput for non-voice data services. As VoLTE would use substantially fewer network resources than say Skype voice, which in turn results in longer estimated device battery life for the subscriber and a more efficient network for the operator.

Telcos in developed markets see VoLTE as both a means to gain competitive advantage over rival service providers and a tool to bolster their brand strength. VoLTE promises operators will be able to fully utilise their IMS infrastructure investments, optimise their spectrum efficiency, and add value to existing voice plans.As the developed markets begin to move towards comprehensive LTE coverage, VoLTE will no doubt become an integral part of network architecture .

Vodafone announced recently that it aims to get both its VoLTE and WiFi calling infrastructures up and running by this summer.With the launch of Wi-Fi Calling and VoLTE, CSPs have a good opportunity to win back their customer's voice usage from OTT VoIP providers. VoLTE and Wi-Fi Calling have a key differentiator compared to OTT players in that they can use a native dialer. Customers make calls and texts as if they are using the cellular network i.e. they use the phone's normal dialer and contacts. There is no need for an app, and your contacts don't need to be using closed user group services to talk to or message each other. As the range of VoLTE and Wi-Fi Calling enabled handsets continues to expand, these advanced voice services give Telcos a better foundation to win back customers' voice usage from OTTs.

WiFi calling should help let users make calls in places where cellular signal is weak or non-existent, sending your voice over a WiFi network instead. It won't reap the same benefits as VoLTE when it comes to power efficiency, but it'll keep people connected in more places.Some of us might be already be familiar with the concept, whether using Skype or carrier's offerings like Three's InTouch app. Either way WiFi calling will negate the need for an additional app, letting you make a take calls from the dialler as you would with a conventional cellular call. Vodafone's offering an approximate launch date of summer 2015, whilst the ability to use VoLTE also depends on the phone's hardware. A number of top new Android handsets along with the latest pair of iPhones support VoLTE, but it'll take a little while for the technology to trickle down to more affordable handset.

ATT in the US is an example of an operator that deployed VoLTE ahead of Wi-Fi Calling.Initially launched in July 2014, AT&T continues to bring VoLTE to more states in the US with an ongoing rollout plan. The operator is expected to launch Wi-Fi Calling by the end of 2015. Many CSPs are starting with VoLTE over Wi-Fi Calling to avail of improved efficiencies over 3G. LTE offers twice the spectral efficiency of 3G/HSPA and more than 6 times the efficiency of the GSM technology. This improved spectral efficiency makes it possible for VoLTE to handle twice as many calls, helping to optimize the use of radio resources and reduce costs. The requirement to migrate circuit switch voice traffic onto VoLTE is also a deciding factor for deploying VoLTE in advance of Wi-Fi Calling.

T-Mobile US made a big splash in the market in late 2014 when they heavily promoted their Wi-Fi Calling service to overcome any coverage issues. To drive adoption of the service, T-Mobile allowed customers to upgrade to a new Wi-Fi capable smartphone at no additional cost. In addition, T-Mobile offered a free proprietary "Cellspot" Wi-Fi router (to all Simple Choice postpaid customers to enhance their in-home coverage). In Q1 2015 T-Mobile reported that 7 million customers were using Wi-Fi Calling. T-Mobile also reported that over 10 percent of voice calls run on its new Voice over LTE technology. The operator expects growth to accelerate significantly and notes handset adoption as the key enabler to success.

When delivering voice services, Telcos need to consider latency, loss, and jitter, which would not normally be associated with many data services. They must ensure that the LTE coverage and Wi-Fi hotspots, accessed by customers to make voice calls, have sufficient coverage to deliver an acceptable Quality of Service (QoS). In relation to VoLTE calls, the challenge lies in being able to dedicate sufficient resources to guarantee the quality of the call. A Policy Control Rules Function (PCRF) is required to define the QoS for the call. This is possible as the CSP has visibility of the QoS from the user's device to the network. The PCRF initiates the creation of a dedicated bearer that carries the voice traffic for the duration of the call. It also guarantees bit-rate for voice data.

T-Mobile Czech Republic has become the first operator in the country to launch Voice over LTE (VoLTE) commercially. VoLTE allows standard IP calls on LTE networks.The first supported device is the Samsung Galaxy S5. T-Mobile is working intensively to offer new models, in cooperation with manufacturers.The biggest advantages of VoLTE is a very short time for call set-up, enhanced quality of call (even compared with HD Voice in 3G) and lower energy consumption per call, or a longer battery life. During the call, there are no restrictions on data connections, e.g. simultaneous use of online navigation and calls when driving a car. When a user leaves the area of LTE coverage, the call is handed over to a 2G or 3G network (function SRVCC). Customers will get notifications requiring them to update their firmware.T-Mobile will subsequently deploy VoLTE automatically to clients on their current telephone numbers. In order to properly monetize VoLTE services, Telcos do need functions of an online charging system sitting in the BSS/OSS so that VoLTE and Wi-Fi calls can be billed in the same way as traditional voice.

Swiss operator Salt launched Wi-Fi Calling in June 2015. Salt states that Wi-Fi Calling turns every Wi-Fi router into a Salt antenna. They reference over 3,000,000 Wi-Fi hotspots available which include those of their competitors (Wi-Fi router Swisscom, Sunrise, UPC Cablecom at home etc. and all public Wi-Fi hotspots). In July 2015, Salt extended their Wi-Fi Calling initiative to include Wi-Fi Calling worldwide, making it one of the first operators in the world to offer this service. The initiative will abolish roaming costs for Salt customers, with all calls and SMSs charged according to the applicable subscription – so its always as if you are making your call in Switzerland.

Ofcourse it makes no difference to customers whether the call is made over the cellular or Wi-Fi network provided the quality is satisfactory.With fierce competition within the industry and poor coverage ranking as a key driver for customer churn, QoS management should and must be at the forefront of VoLTE and Wi-Fi Calling roll-out strategies for all Telcos.The bane of OTT VOIP is poor QoS and dropped calls..and that is why VoLTE + WiFi Calling is a golden opportunity for Telcos to claw back some of what they have lost to OTT's.

LTE Monetisation is predicated on many factors that play at the same time in a subtle equation that responds to certain levers: dynamic data pricing / billing , appropriate device porfolio , Enterprise Verticals , wholesaling bandwidth , QoE and CRM , business model innovation , Cloud , OTT partnerships...and most of all : very clever thinking !!

---◆---

Telco Challenge # 5 : The Quest for $$ in the OTT World

For some time now the OTT's have been viewed as the anti-carrier Antichrist. Just recently the CEO of one South Africa's leading MNO's told a tech news agency that over-the-top (OTT) services like WhatsApp and Skype were unfairly benefiting from his company's costly infrastructure. He warned that his network was not prepared to spend billions on its network just for the OTTs to have a "free ride". Another Telco was quick to declare that OTT services were "skimming" its voice revenues. Without these extra revenues, the company claimed, it cannot afford to provide telecommunications to poor rural areas, where it runs at a loss....boohoo...Did these guys just wake up because the OTT invasion began 5 years ago ??

"Why are the next two billion not on the internet?. The reason is not because they don't have any money, it's because they don't know the value of having a data plan or the services they can access." (Mark Zuckerberg, founder and CEO of Facebook)

It all seems so unfair that OTT players make their money by loading more and more traffic on the operators' networks, thus causing the operator to pay for network capacity upgrades to facilitate the OTTs' increasing VoIP and IM traffic that was the root cause for decreases in the operators' core voice and SMS revenues. But if truth be told circuit switched voice and SMS are on the way out, and as operators move towards becoming fully IP based digital services providers. So some clever Telcos are entering into collaborative agreements with OTTs in order to generate new revenues and provide a deeper range of services to their customers.... instead of making dire predictions and pathetic threats !!

While working with OTTs is still at an early stage for many operators the $$ opportunity is to not view OTTs as a threat but to use them to build more comprehensive offers that customers want and help to build loyalty. Instant messaging is on the rise and texting is starting to see its first decreases in usage since it was launched. In August 2014 research firm Deloitte reported that the average person sends seven text messages a day, compared to 46 instant messages. SMS texting is forecast to fall from 145 billion to 140 billion by end of 2014. According to the survey of UK consumers almost a quarter of smartphone owners use five or more messaging apps.

"Everybody worries about being a dumb pipe, and whether revenues will be able to support network investment that we need to make.Any other industry would be excited and highly optimistic given the strong demand in growth for their core services. However, the big problem we have as an industry is we have been unable to monetise this increased demand "

(Chua Sock Koong , group CEO at SingTel)

There are many operators who have started selling application service passes which provide a low cost means of using an OTT service from a smartphone. For example, in Kuwait, Ooredoo offers a WhatsApp service for only 750fls (2 euro) a month. This is a good example of offering a low cost service limited only to a specific application to generate a new income stream and encourage mobile data adoption. Ooredoo promote this service as an alternative to trying to find a free Wi-Fi zone. As well as offering low price services for one app, operators are also zero rating popular services, such as Facebook or Twitter in order to get customers using their services.

In June 2014 TMF ran a survey of 67 operators looking at uses of real-time convergent charging and policy management. In this survey 56% of operators said they are offering zero rated deals, under which the use of services such as Facebook and WhatsApp don't count against customers' data allowances.

Operators can look to work with OTTs in order to develop new revenue streams to replace the traditional texting and voice revenues. It's not a case of operators 'versus' OTTs. Operators have some key assets that can make them very attractive to OTTs as a route to market and develop win-win partnerships. These include but not limited to :

• Established customer relationship often built up over years

• Established financial relationship – regular billing and subscription models, and prepayment for services

• Flexible charging, pricing and billing options – ability to provide innovative pricing plans which can help add value to the OTT's services

• Ability to prioritize the delivery of certain services – e.g. in a congested network operators can apply traffic prioritization

• Ability to differentiate service delivery – e.g. tiered service delivery: supply certain video traffic over LTE and offload others to slower Wi-Fi networks to avoid network congestion

• Use big data and customer analytics (e.g. cell site location, customer behavior) to upsell more relevant services and finely tuned advertising

E Plus in Germany and China Unicom have launched prepaid SIM cards that lead with OTT services. The WhatsApp SIM card from E Plus and the WeChat SIM card from China Unicom offer mobile voice minutes, a data tariff and zero rated OTT IM services. Looking at the E Plus example , which was launched in April 2014, this offers a prepaid card which costs €10 / month and the main message is WhatsApp is always free, even when the subscriber runs out of credit (in the 30 day period). This is a major benefit over other networks who charge for the data required for WhatsApp (which is installed on 90% of all smartphones in Germany, giving a potential user base of more than 30 million people). By working with this OTT E Plus can leverage the popularity of WhatsApp to attract new customers and open new revenue streams.

China Unicom has worked with the OTT IM service WeChat to offer a WeChat China Unicom SIM card that offers additional services over the traditional WeChat mobile IM service. These additional services include increased group chat limits and extra free icons. By differentiating the WeChat service available with the China Unicom WeChat SIM there is another clear incentive for WeChat mobile customers to switch to China Unicom. This service was launched in August 2013 in Guangdong and it was reported that in one month it gained 1 million customers.

The most recognizable players in streaming music , Spotify, Beats Music, Deezer and Napster have all partnered with mobile operators. his may help explain Spotify's increase in paying customers from 6M in March 2013 to 10m in May 2014. Deezer has probably been the second most prolific in terms of mobile partnerships and they say their paying subscribers increased by 4M to 5M between May and November in 2013. Sprint have added Spotify to their Framily (friends and family) plan. All Family customers get a six-month trial of Spotify and they will get Spotify at the discounted rate of $7.99 a month. For family groups of 6 to 10 members, the rate falls to $4.99 a month.

Telecoms operators have a choice - they can collaborate or compete with OTT service providers. Operators in the Middle East are taking both approaches. While incumbent operators in Saudia Arabia (STC), the United Arab Emirates (Etisalat) and Qatar (Ooredoo) are competing with OTT providers by either extending their IP TV services to OTT platform or by emulating popular OTT services, newer entrants to the market such as Saudi Arabia's Mobily and Nawras in Oman have collaborated with WhatsApp to launch specific packages. Because launching in-house OTT services or partnering with OTT players can be a double-edged sword, taking the precautionary step of adapting the services to the operator within the competitive context is highly recommended.

The recent interconnect deals with AT&T; Verizon and Comcast that Netflix has struck illustrates that in order to deliver video at the speed and quality that the content providers (and consumers) want then telecoms operators and customers cannot be expected to pick up the delivery bill all the time. Perhaps this illustrates that content providers are recognizing (and will pay for) differentiated service delivery. With over 50 million subscribers Netflix is enjoying substantial growth and has established payment of a monthly subscription fee. With viewing behavior changing and more consumers watching videos and TV shows on tablets and smartphones, companies like Netflix are seeing the potential for working with mobile operators (and vice versa). If operators are looking at partnerships with video / TV content providers then the first thing they need is real-time data collection, rating and notification systems.

Rather than just have all video traffic on a LTE network an operator could offer options to offload onto available Wi-Fi at a lower (or no) cost. This is done by having on device ANDSF (Access Network and Discovery Function) software which enables operator controlled offload by assisting devices to discover access networks in their vicinity (e.g. Wi-Fi) and provide rules to prioritize and manage connection to all networks. This allows operators to dynamically control and define preferences – that is how, where, when and for what purpose a device can use a certain radio access technology – e.g. under what conditions is traffic offloaded to Wi-Fi.

Having data delivery charges picked up, or subsidized, by the content provider or by advertisers could be a possible way to take the sting out of data heavy content charges. The delivery of media content has been traditionally subsidized (television, cinema, newspapers, radio) and it could be argued that this will not change for digital content delivered on mobile networks. If a customer was to watch a 30 minute TV show each day / month on a LTE network then the data charges would be around US$70 / month (for 4GB). Most customers would eschew this. If part of the cost was picked up by advertisers then maybe the retail costs could come down, with the resultant increase in active subscribers, content usage, and advertising revenues.

However the systems required to support much more diverse and dynamic business will need to be very different from the traditional legacy BSS that were designed to collect, rate & bill for circuit switched voice calls and SMS text messages in batch mode. Legacy BSS systems are the perfect recipe for bill shock and some very angry customers besides being inflexible when it come to dynamic service creation , delivery and monetisation.

Many telecoms companies are now pursuing multi-play strategies in which they offer a number of services, including mobile and fixed voice services, broadband and pay-TV. Vodafone has been active in this market with several high profile acquisitions in Europe, while the deal between BT and EE in the UK will also create a quadplay powerhouse. CCS Insight research suggests that consumers are interested in signing up to companies offering this range of communication and media services if their offering is good value. The appetite for multiplay services may also be driven by people owning an increasing number of devices.

" Operators are competitively disadvantaged with over-the-top players, owing to heavy infrastructure investment. Regulators should apply the principle of "same service, same rules" to players of every hue, and that consumers should have a "portable digital life" where it's just as easy to switch digital ecosystems as it is networks.

(Caesar Alierta , CEO Telefonica)

Telcos today are facing margin pressures through more intense competition, ARPU erosion, customer churn and cost issues. While designing new business models, Telcos can leverage their network capabilities, such as mobility, messaging, location, presence, profile and call control, and combine these with internet-style services such as social networking, search, advertising, direct marketing and mapping, thereby enabling richer, more compelling and more personalised services than the Internet players can offer.

Furthermore, by exposing these capabilities in a secure, controlled and automated manner, Telcos can generate revenues from selling service enablers, as well as their own services, allowing them to fully exploit their network assets.

In light of the above Telco Execs need to understand that :

1. The design and bundling of applications, content and devices to generate revenue from broadband networks is based upon a deeper understanding of the customer's data consumption habits

2. The business and technical logic underlying services delivery platforms because telecoms networks have evolved from voice-centric "legacy" technologies such as SS7 and IN towards data and multimedia-centric technologies based on IP, such SIP , Daimeter and IMS

3. The critical role of converged billing and CRM engines and how to convert BSS/OSS into revenue generating assets and the need to introduce attractive, profitable new services to subscribers with minimum time-to-revenue while controlling costs

Telcos can greatly benefit from implementing convergent customer care and billing systems because investing in a new stovepipe billing system for each type of service is an expensive and obviously sub-optimal proposition. The systems would help them bring new services to the market quickly, enabling them to improve customer loyalty and reduce customer churn, especially in an environment, where customers jump from provider to provider to get the best deals.

Big data has been a headline theme in the technology and mobile space for some time.Telcos all over the globe are seeing an unprecedented rise in volume, variety and velocity of information ("big data") due to next generation mobile network rollouts, increased use of smart phones and rise of social media. Telco operators have historically focused on managing the network with little visibility to the impact it has on the customer's experience. Which means the operator was forced to work with snapshots of network data in fragmented views or at a summary level in order to plan network capacity or provide information to customer care and marketing about customer transactions until now !!

Big Data technologies, and in particular their analytics abilities, offer a multitude of benefits to telecom companies including improved subscriber experience, building and maintaining smarter networks, reducing churn, and generation of new revenue streams.

Mind commerce, expects the Big Data driven telecom analytics market to grow at a CAGR of nearly 50% between 2014 and 2019. By the end of 2019, the market will eventually account for $5.4 Billion in annual revenue.

Mobile commerce is one particular area where operators and service providers can potentially deliver tangible benefits from the application of big data analytics. The growth of m-commerce is creating large amounts of information on consumer behaviour and choices, which can be used to offer more personalised services and offers. SK Planet (Division of SK Telecom) have stated that "our Cash Bag m-commerce portal should generate $9.3 billion in revenues this year, and by using big data analysis we can provide customers with a much improved experience, and not based simply on offering the lowest price."

Big data analytics solutions enable service providers to analyze real-time location data over time for opt-in subscribers to understand subscribe lifestyle. Combining lifestyle and mobile profiles with subscriber usage and digital behavior allows service provider to create targeted offers for opt-in subscribers. With a majority of subscribers using smart phones to access data services as well as voice, mobile network operators are seeing explosive growth in traffic levels across their networks. In addition, the mobile network operator environment is fiercely competitive, with the ability to attract, retain and grow valuable subscribers being key. Increasingly, the provision of high quality customer care is an important component in the marketing mix and in retaining subscribers.

The growth of connected devices, particularly in areas such as the home or in the car, presents new opportunities but also challenges for operators and other ecosystem players. Users may be willing to share data with service providers but on the basis that the data is used securely. This year the GSM industry introduced a standardised mobile identity solution that aims to become the de facto single sign-on tool that consumers could rely on to authenticate themselves in both online and offline environments. This initiative is set to stimulate adoption of mobile services that rely on absolute confidentiality, such as healthcare, government and banking.

The Gurus at Strategy & believe that many types of data are potentially available to operators — and certain sets of data might be combined to open up new business opportunities in areas such as campaign marketing and fraud prevention. Operators could generate more accurate and personalized offer recommendations for existing individual subscribers by combining internal structured data, such as how and where each subscriber uses his or her phone, with external unstructured or semi-structured data from social media platforms (for example, Facebook and Twitter).

By correlating internal location, usage, and account data with external sources such as credit reports, operators could significantly increase the detection of fraudulent activity such as looping or call forwarding on hacked PBXs (private branch exchanges), or fraud involving the swapping of SIM cards, and improve the overall accuracy and efficiency of their efforts to recognize patterns of fraudulent behavior.

Imagine having the best of both worlds ? Having the tools to analyze the growing amount of data and content your business is generating, and also finding ways to make it profitable. If you are astute then this deluge of data isn't a threat; it's a serious opportunity to take your telecom business in a new, exciting, and yes, profitable direction !!

In its most basic form, Enterprise 2.0 is about communication. When information is free, people can get more feedback and input (collaborate), react more quickly (agility), and make better decisions. This is the opportunity inherent in Enterprise 2.0: a more efficient, productive and intelligent workforce.The current crop of Collaborative solutions focus around unified communications (instant messaging, web conferencing and VoIP for example), working in teams, sharing documentation and knowledge, working with (self-service) portals and working with social collaboration tools.

Organizations are beginning to take advantage of social collaboration aspects like communities, blogging and wikis to connect with external parties like partners, customers and local government. A survey done by McKinsey & Company showed that companies that benefit most from B2C/B2B collaboration are:

- Networked organizations;
- Business to business organizations;
- Big companies (> $1 billion revenue);
- International companies;
- Decentralized organizations.

According to Industry experts there are three fundamental ingredients to be successful with E20/Social Business (or any major corporate initiative): Adequate resources/budget, organizational commitment, and a business problem to solve. Missing any of these greatly slows down and/or blunts the outcome of the effort. The top challenge is culture change. You can drop social technology into any organization, but you can't suddenly expect that employees will adopt the way that social media works or that business processes or traditions will automatically change.

Social is a new way of operating (observable work, openly participative processes, co-creation) and this requires conscious effort to change our thinking and the way we function. Other top challenges include enterprise apps with overlapping features (e-mail, CMS/DMS, IM, unified communication, enterprise microblogs, customer forums, CRM, etc.), underinvestment in community management, and lack of executive understanding or buy-in.

Web 2.0 is the term for web-based tools and services that allow for – and even improve with – user participation. The most well-known examples of this technology are found in sites like YouTube , Facebook , Wikipedia and Amazon, where users to find and connect to what they are looking. Social media tools like blogs and microblogs (Twitter) opened up the world of media and publishing to anyone with an internet connection – or a smartphone.

Social network tools help staff find the right individual or group of people.Tagging and rating provide a straightforward way to find content and make judgments about what to look at. Blogs and wikis are natural collaboration and communication platforms.Giving employees the freedom to speak their mind and voice ideas is required for there to be a harnessing of collective intelligence

One of the biggest car lease firms worldwide had a clear vision on the use of social collaboration, both internally and B2C/B2B. In this vision it outlines the many strategies it intends to use for social collaboration.

* Launching an employee community
* Engaging in social recruitment
* Social software enabled car remarketing
* Launching fleet management communities
* Social software enabled car quotations
* Launching driver and supplier community
* Launching a supplier community
* Conducting online reputation management

A Global management consulting, technology services and outsourcing company implemented off the shelf social media platforms technology to introduce knowledge sharing communities and social networks. Blogs and wikis function as collaboration tools, and as such, they have uses mainly in sharing "unstructured" information associated with ad hoc or ongoing projects and processes, but not for "structured informational" retrieval.

However, Shell has started converting its official documentation to wikis, because this enables that company to make documentation updates available in real time and allows non-editors to contribute to the documentation. In this process Shell restructures the paper documents to a set of on-line wiki pages. Their key challenge was to get key stakeholders aware of how social computing can solve business problems and be integrated into business processes. The business case was based on the following metrics:

• Finding people and identifying experts;
• Finding information;
• Reducing the need for travel;
• Speed up the decision making process.

Current Telco mainstream offerings to the Enterprise market are based around capacity and hosted services, sometimes complemented by IT outsourcing projects. Mass-market consumers and Enterprise customers alike are increasingly demanding rich, portable, personalised, access and device-independent services from their Telco Service Providers.Telco 2.0 and Web 2.0 components creates more value to the Enterprise . For example Telco resources can be embedded with the Enterprise applications to identify the real time location and distribution of a service engineer's customers (using Google Maps and Location feeds) to view the geography of the area covered.

In the UK, BT (British Telecom) has become one of the country's strongest proponents of enterprise 2.0. The company has introduced a raft of social media tools, including a huge Wikipedia-style database called BTpedia, a central blogging tool, a podcasting tool, project collaboration software and enterprise social networking.

With SDP / IMS platforms Telco customers can become part of the social networking phenomena by complementing these content based sites with telecom capabilities such as anonymous calling (whisper calls) and real time updates showing the physical location of friends and contacts within the community group. Whilst Telecom Web Services standard will revolutionise Telco service offerings to both the Large Enterprise and the SMB market, it is important not to overlook the benefits of being able to offer fully hosted services.

These include "Virtual PBXs" and "Virtual Contact Centres", plus a suite of complementary services to customer's own installed platforms such as "Mobile PBX Extensions", "Multi-line" services for handsets and of course "Voice Call Continuity" to provide Enterprise roaming in WiFi hotspots.The availability of Enterprise 2.0 tool combined with high speed networks smartphones and cloud computing will unlock fresh new revenue streams for agile Telcos and CSP's in the Enterprise / SMB markets.

We are on the cusp of a management revolution that is likely to be as profound and unsettling as the one that gave birth to the modern industrial age. Driven by the emergence of powerful new collaborative technologies, this transformation will radically reshape the nature of work, the boundaries of the enterprise, and the responsibilites of business leaders.

The development of electronic signature in mobile devices is an essential issue for the advance and expansion of the mobile electronic commerce since it provides security and trust in the system. E-signatures provide security for the transactions with authenticity and integrity characteristics that make non-repudiation of the transactions possible. In many countries, such as Estonia, Germany, Singapore and Hong Kong, it has become a key element of e-government through its "utilization of wireless and mobile technology, services, applications and devices for improving benefits to citizens, businesses and government units .Driven by the growing surge for mobile interactions, mobile commerce and online digital purchasing, carriers worldwide are investing in mobile identity infrastructure as an economically efficient solution fraud detection/prevention and identity theft issues.

Many different technologies and infrastructures have been developed with the aim of implementing mobile signature processes. Some are based on the SIM card. Others work over the middleware of the mobile device and cryptographic providers. Finally, there are already some frameworks which are independent of specific mobile device technologies and make mobile signatures available to application providers.Mobile signature solutions can only work on compatible SIM cards, that match the WPKI specifications in terms of security and capacity, and contain a SIM Toolkit application capable of performing signatures. The PKI system associates the public key counterpart of the secret key held at the secure device with a set of attributes contained in a structure called digital certificate. The choice of the registration procedure details during the definition of the attributes included in this digital certificate can be used to produce different levels of identity assurance.

A solution must also be implemented on the operator side to manage signature requests. If the access control secret was entered correctly, the device is approved with access to secret data containing for example RSA private key.Security is guaranteed by cryptographic systems (e.g. SHA1) and on-board key generation. The service is only made available on EAL4+ certified SIM cards which provide a high level of security. Legal compliance is ensured by a country specific Electronic Signature Law that gives electronic signatures the same authentication level as wet signature as long as they rely on a "qualified certificate". Qualified certificates are defined by the ETSI Standards4 and a directive by the EU Commissions as certificates that are issued by an authorised Certificate Authority following face-to-face verification of both the user and government issued photographic identification.

According to GSMA , Turkey and Turkcell was the global first in launching a mobile signature service. The idea behind Turkcell MobilImza was to offer a remote way to complete transactions equivalent to an "original" signature on a hard copy – making it possible to sign documents and authenticate oneself via a mobile phone,in a way that is legally approved, secure, easy and convenient.Their Mobile signature services are easy to use, since they don't require any software installation. The certificate is activated Over-The-Air once the user has subscribed to the service. Signature requests then automatically pop-up on the user's phone each time he requests access to secure services. Once the user has entered his PIN, the signature is sent to the service provider, who checks its validity and grants access to the service.

Turkcell Mobile Signature can be used in all transactions, except for the ones that require a ceremony and the witnessing of a third party such as marriage or buying a deed, that require a signature such as private affairs, public affairs, and banking affairs. For example, EFTs can be carried out over internet banking with mobile signatures. Turkcell has imported e-signature technology to the mobile realm and it contributes to the e-Turkey transformation by carrying all transactions that require signature to the virtual realm where you will not need new readers or smart cards.

It is often the case that service providers are reticent about adopting mobile signature solutions if there is not a large installed base of users, and users are not enthusiastic about services that are not backed by multiple service providers. This leads to a stand-off that can often threaten the commercial success of mobile signature services. Initially, Turkcell's project was supported by the five main Turkish banks, which together pushed for the government to adapt the electronic signature law. This collaboration helped drive adoption since banks offered customers pre-registration at their branches, and then sent the forms to Turkcell. The banks also promoted the use of mobile signature through marketing campaigns.

The initial business model for Turkcell Mobilimza was a pay-per-use model. The service was free to subscribe to, and users had to pay a fee each time they used the signature service. The idea was that the cost of the certificate would be covered after a certain number of transactions, and then profit would be generated by extra usage. But this model relied on consistent levels of usage from subscribers. However a significant proportion of non-active users made this model unsustainable. Therefore this business model was replaced by two complementary approaches:

■ Monthly subscription: subscribers pay 5 Turkish Liras for an unlimited number of signatures
■ Price per signature: service providers pay a small fee per transaction. Public enterprises and educational institutions are not required to pay this fee, because of their public service orientation. It is anticipated that service providers who actively promote mobile digital signature will also enjoy a waiver of this fee.

In Europe the Mobile ID program in Moldova is a government-led project that is being deployed in partnership with mobile network operators. It is designed to offer citizens the speed, privacy, convenience and transparency of digital access to numerous government services and information for citizens, including online applications and copies of official documents. Their selected UICC-based solution is compatible with all types of mobile telephones, whether feature phones or smart phones. The application allows citizens to confirm their identity and sign documents directly from their mobile phone, by entering a unique user-selectable PIN code. A Mobile ID solution is responsible for the entire life cycle management, from user registration to verification of mobile digital signatures, and connection to the Certificate Authority body and e-government portals.

Lattelecom offers the Mobile ID service platform to Latvian service providers and mobile operators. The Mobile ID users are able to securely sign in to online services and sign documents and transactions, simply by using their mobile phone. As mobile phones are typically always at hand, a legally binding digital signature can be done regardless of time or place. Lattelecom launched the service with nine service providers, including Latvijas Krājbanka bank, Riga city council, Lattelecom's and Latvijas Mobilais Telefons' (LMT) online customer service, along with local enterprises and universities. Lattelecom handles user and transaction validation, making it easier for other service providers to join in. Lattelecom offers the Mobile ID service to third-party service providers in a variety of industries, including other Latvian mobile operators.

The ability to leverage network assets, such as the Subscriber Database Management (SDM) system, and the potential for incremental revenue from third-parties such as credit bureaus, banks, and credit card companies, makes mobile identity a high priority service for carriers worldwide. Mind commerce thus expects that mobile identity infrastructure market will grow at a CAGR of nearly 17% over the next five years eventually accounting for nearly USD 12 Billion in revenues by the end of 2019.

If you don't like having your credit card tied to your Google Play account and live in Singapore, here's the good news: you can now link it to your monthly phone bill (or prepaid balance) from August 15 onwards if you're a SingTel subscriber. SingTel has announced a partnership with Google to be the first in Southeast Asia to offer carrier billing services. This means you can buy apps or in-app purchases and the amount will be charged to your mobile line instead . The company also mentioned that this will be rolled out to Optus in Australia, and companies where the company either has shares in or owns in Thailand , Philippines, India , Bangla Desh and Indonesia.

Aaah yes : welcome to the world of Direct Carrier Billing. Finally the Telcos have woken up to this latent and gigantic opportunity to capitalise on the App store phenomena. Telcos have always coveted a piece of the $770 billion mobile payments market. Direct Carrier Billing helps operators grow new revenue streams by leveraging their key asset: the billing relationship with consumers. Many consumers prefer carrier billing over credit and debit cards, as it simplifies the purchase experience and avoids the need to disclose card details to small or unknown merchants.

Content paid for via direct operator billing will come to a represent a $13 billion revenue opportunity by 2017, according to Juniper Research, compared to $2.3 billion in 2012. The research indicates that Google Play, Nokia Store, Blackberry App World and Windows Phone Store – all of which offer Direct Carrier Billing in a number of markets – accounted for approximately 48% of all app downloads in 2012. And for you NFC afficiandos : while trials have demonstrated extremely positive user responses to NFC– given the scale of the marketing/educational challenge facing MNOs and other NFC stakeholders – Direct Carrier Billing represents a greater and easier monetisation option in the short term !!

Direct Carrier Billing allows storefronts to enable payment amongst a far wider and diverse user base, both in developed and developing markets. In the latter case, bank account and credit card ownership is often extremely low; in the former, it provides a billing option to the prepaid sector and younger demographics. Furthermore, it enables few-click purchases, thereby making it a particularly attractive option for impulse purchases. Conversion rates are much higher than many other payment means, due to ease of : for example, the Google Play transactions processed on carrier billing in 2012 in the US grew by 3x in comparison with 2011. Nokia CEO stated that in cases where Nokia teamed with mobile operators to offer carrier billing, consumers were 5x more likely to complete an app store purchase than if the app store only offered credit card purchasing.

There is significant opportunity for Direct Carrier Billing to be utilised for 'real world' purchases in the lower value, higher volume area, both in terms of ad-hoc purchases (e.g. books, flowers) and more regular purchases (e.g. petrol). By adding ticketing applications and services to a mobile phone a customer could be less likely to replace their mobile operator with a new one; customer loyalty should increase as a result of mobile payments. With the addition of analytics of the BSS and Big Data, the operator can add far more value to mobile commerce. This is beneficial both from the perspective of the MNO – allowing it to personalise content discovery based on analysis of on-going individual consumer behaviour patterns – and for third parties such as advertisers and retailers. In several major cities in Central & Eastern Europe there are mobile ticketing schemes for trams and buses which are well established and see multiple millions of tickets sold annually. Examples include Prague, Warsaw, Bucharest, Bratislava, Zilina, Kosice and Tallinn

Telefonica Digital, along with the Telenor Group is now offering Carrier Billing to a combined customer base of 400 million. Impressive as that figure is – and i do not know how many of those customers have smartphones or tablets – the best figure is that revenues from payments have increased by 300 percent. Credit card and PayPal transactions have suffered as a result. In Latin America that figure is closer to 1,000 percent because of the low penetration of credit cards. In fact the new Firefox Marketplace, aimed at that market, is the first application to have Carrier Billing 'baked in.' Samsung has agreed to let customers on Telefónica's network charge their apps and content to their phone bill or take the payment out of prepaid credit, rather than having to use a credit card.

Typically new Carrier billing platforms are Cloud based gateway / middleware coded solutions dishing out SaaS . They enable Telcos to onboard and manage the largest set of app stores, merchants and aggregators, and process transactions for any type of goods – digital, remote, or physical. In addition, they make it easy to settle against any payment method (postpaid bill, pre-paid balance, wallet, or card) and quickly capture new and emerging revenue streams. These platforms include automated workflows and business processes that maximize operational efficiencies. While the cloud-based service delivery model offers low CAPEX requirements, the revenue-share pricing model on offer ensures that OPEX is tied to service providers' success and payments revenue.

We're going to be seeing more of these carrier billing arrangements in the future. Not only does it mean more apps and content will be sold, benefiting their developers, but it also means the Telcos themselves aren't shut out of the value chain. This shift towards giving Telcos a slice of the Apps Store pie is a good thing – not because the proactive operators deserve it because it rewards them for the investment in expensive network assets that enable the Apps and Mobile payment economy.

The emergence of voice over Internet protocol for mobile devices, or mobile VOIP, has the potential to transform the mobile device and telecommunications industries. The arrival of low latency LTE technology will finally resolve the latency problem in VOIP even as smart phones with integrated wifi continue to attract new customers to mobile service providers.

The capabilities of smart phones and a departure from the traditional contract structure for mobile wireless plans means that support for mobile VoIP is growing. The latest technology innovations such as low powered single chip radios with integrated VOIP protocol stacks, devices with integrated multimedia terminal adaptors and large scale, centralized Wi-Fi infrastructure promise to drive mass adoption of mobile VOIP.

The mobile VoIP market is expected to be worth $32.2 billion by 2013 and by 2019, moreover half of all mobile calls will be made over all-IP networks, according to recent industry reports. VoIP providers are also integrating Facebook into the desktop client. The social media giant has recognized the potential and is now offering its own mobile VoIP client for iOS. This move is just another indication of the role smartphones and tablets play in the overall mobile market, and the need for mobile VoIP providers to leverage strong partnerships with device providers.

For many enterprises , the optimum solution may be a combination of Wi-Fi and cellular. A Wi-Fi/cellular roaming solution requires dual-mode handsets that support both VoWiFi and cellular—and a network gateway. The gateway manages access and handoff and connects to both a mobile switching center for cellular calls and a data network for WLAN calls. As people move within range of a wireless access point, the gateway authorizes access and delivers both voice and data network services over the WLAN. When people move outside of coverage of the current WLAN, the gateway seamlessly switches control over to another WLAN or a cellular network if another WLAN is not available.

VoIP over a wireless LAN can provide easy internal calling for corporations, educational campuses, hospitals, hotels, government buildings, and multiple-tenant units such as dorms, with the ability to roam freely and advanced calling features such as voicemail and caller ID. Users can also use the LAN's Internet connection and an account with a VoIP provider to make calls outside the site, including domestic long distance and international calls, often at no extra charge.

We can now envision a scenario with interconnected fixed network environments and mobile network environments, in any combination, including 3G on the mobile side, and with fixed broadband, cable, and ISP environments on the other. When making the transition to IP, Telcos must keep in mind subscriber demand for seamless functionality and consistency across multiple service provider networks and types.

Standards based Mobile VoIP solutions offer simplified converged services between mobile and IP networks that reduce coverage disparities and operations costs to deliver competitive subscriber prices, consolidated billing, and subscriber loyalty. Mobile VoIP solutions utilize Network Convergence Gateways that leverage standards based on SS7 signalling enabling voice and data calls on any SIP-based client.

The coexistence of wireless heterogeneous networks has been widely recognized, and it has become more common that new mobile devices get equipped with multiple and heterogeneous wireless interfaces. Furthermore, the recent advances in software-defined and cognitive radio technologies including the availability of TV white space spectrum promise even more diversity and heterogeneity. This presents lot of opportunities and challenges for mobile wireless networking. Environment cognizance, spectrum-aware mobility management, and vertical handoff thus become critical components in the Mobile VOIP solution space as does correct network design.

The mobile industry developed a standard for all-IP operations called Internet Protocol Multimedia Subsystem (IMS). The IMS standard promises to allow service providers to manage a variety of services that will be delivered via IP over any network type – including the mobile network's packet switched domain (GPRS, 3G , 4 G). With IMS, service providers will use IP to transport both bearer traffic and Session Initiation Protocol (SIP) for signalling.

Any IMS strategy must include a solid plan for supporting subscription-based, usage-based and tiered billing.The ideal IMS solution must integrate smoothly with the OSS because the picture grows even more complex with content-based billing, context based billing and differentiated billing by QoS. Disappearing are the days of only flat-rate billing; transaction-based billing on application usage and subscription profiles is a likely future reality. You need an IMS solution that can reach across domains, for three big reasons:

1 : Quality of user experience — Your subscribers are going from one domain to another in real life. You need to provide a superior end-user experience and convenience wherever they are, on the same device or different devices.

2: Speed to revenue — When services can be delivered across access types, you'll see faster acceptance among a larger community of interest.

3 : Future-readiness — You need the flexibility to address new and evolved business models two to three years into the future, even if you don't know where your company will be.

Key Benefits of Mobile VOIP include but not limited to :

• Communications service providers can capture wireline Minutes of Use (MOU)
• Improves quality of mobile voice and data services in homes, offices and public access venues
• Extends the subscribers single identity to the fixed-wireline venues and strengthens the brand to create a "sticky" service
• Open architecture yields best performance at lowest infrastructure cost
• Extends mobile footprint via unlicensed spectrum thus reducing capital expenditures on mobile base station construction and infrastructure
• Leverages the economics and prevalence of Wi-Fi and SIP
• Maximized existing and future infrastructure investments by deploying pre-IMS (IP Multimedia Subsystem) network elements today.

02 UK is launching its Tu Go app the operator said will enable its users to make and receive calls, texts and voicemail via the Internet using their existing telephone number.The service, available on all Apple and Android devices, is free to download for 02 contract customers, with the calls and texts taken from their existing bundle.The aim of Tu Go is to free customers from being locked to a single handset : Customers can now take their mobile number wherever they like, even away from their mobiles !! 02 UK

customers can be logged into the Tu Go service on up to five devices at once, said the operator.

Incoming calls will ring all logged-in devices, including handsets using SIM cards associated with different networks and Internet-enabled gadgets such as iPods. Only available to O2 UK's postpaid customers, it is a cloud-based telephony service, allowing the user to register multiple devices and make and receive calls and messages from all of these as if from their telephone number. Any usage comes from the user's postpaid inclusive bundle.

TuGo can therefore be used regardless of physical location over Wi-Fi using the user's home contract. This also makes it an FMC solution, because it will work indoors on Wi-Fi at places where mobile coverage is poor.Significantly, the integration with native communications services (telephony and SMS) means that users are not restricted to communicating with other TuGo customers. Exchanging calls and messages with users of basic services works well, with information (e.g. caller ID and dialed number) shared between the native dialer app and the TuGo service. All communication is organized in threaded timelines, and is displayed regardless of the device used.

Facebook paid a whopping $19 billion to acquire WhatsApp. Currently a messaging application, WhatsApp is looking to implement voice over internet protocol (VoIP) by June of this year. A prime example of the growth and prosperity of VoIP in the business market today, WhatsApp already has 465 million users – all of whom will be able to make VoIP calls after WhatsApp releases an update by the second quarter of 2014.

The world is poised for Mobile VOIP revolution with the arrival of ample submarine bandwidth , the continued expansion of terrestrial fibre optic , low cost dual mode smartphones on prepaid and budding LTE networks. So the big question is : Are the Telcos going to profit from Mobile VOIP by offering carrier grade solutions to consumers and enterprises or will they let the OTT players to pile in and eat their lunch as usual ???

To compete head on with OTT players Singtel launch of Wavee, a next-generation IP-based communications app that allows users to make high-quality voice and video calls, and send instant messages. Singtel is the first mobile operator in Southeast Asia to launch such a service.Available worldwide, the app is free for anyone to download and use from the Android Google Play and Apple App Store, and unlike many other messaging apps, does not allow advertising.

Singtel's VP for Consumer Marketing said, "People are looking for richer, more seamless and personal ways of communicating. With Wavee, we are giving our customers a holistic, end-to-end service."

As network conditions can vary between locations, Wavee is able to detect the strength of each user's network and optimise the call quality for the best experience. Video calls will soon also be equipped with this ability. As the app is hosted on a local telco network, customers are provided with an additional layer of quality assurance. Users enjoy unlimited voice calls as long as they are on a data or WiFi connection.With Wavee, users can:

- Make high quality voice, VoIP and video calls, send instant messages and SMSes
- Set up chat groups
- Attach photos, videos, sketches, audio files and documents to messages
- Share a location– either their current location or a place on the map
- Multi-task - sending messages and files while on a voice call

Android users have the option to integrate SMSes into their Wavee message inbox – choosing Wavee as their default messaging app directs all SMSes to the app, reducing the need to toggle between the two. Singtel iPhone users will soon be able to enjoy this feature, enabling friends on both platforms to chat using one central inbox.

To up the fun quotient, Wavee features a wide range of specially designed, locally themed stickers, including one with cheeky army-related references. Users can send their friends and family a sketch, choose a photo, file or colour as the background, and easily overlay a border, sticker, text or a freehand sketch.

"Wavee is the first of its kind to combine all these features on one app, with a distinctively local touch," said Singtel. "For instance, it allows users to send files and messages while on a call, which most other similar apps do not."

At launch, both the caller and receiver must have the app to enjoy the free calls which can be placed via WiFi or 3G/4G networks. A new feature that will be available in the second half of 2015 will allow calls from Wavee to any local number.The range of stickers will also soon increase with many more locally themed sticker sets. Singtel is inviting local designers and design students to collaborate on these stickers

The GSMA believes that mobile operators, through their evolving capabilities in creating meaningful connections between people, organisations and ultimately systems, can have a dramatic impact on the healthcare industry in improving access, reach and quality. Ministerial sessions on m health certainly highlighted that developed countries are under intense pressure to reduce the cost of healthcare, whereas developing countries have a greater need to deliver life-saving services more broadly.

The session deliberated on several key questions : What role can mobile play in delivering future healthcare worldwide? How can governments embrace the significant opportunity to improve quality of life for individuals, and increase access and efficiency in health services for its citizens? What are the emerging best practices from mHealth implementations. Believe it or not TB is a largely curable disease but requires six months of diligent adherence to the medication regime. mHealth could help control TB mortalities by ensuring treatment compliance through simple SMS reminders. Delivering mobile-assisted awareness to pregnant mothers and traditional birth attendants could reduce perinatal and maternal mortality by 30%.

In some developing countries, where there is less or totally non existing technical infrastructure and eHealth systems installed in hospitals, so the focus will be on basic health data collection, basic ICT infrastructure such as connectivity, and health access. Clinical adoption issues (relative to the developed world) will be lower, although the degree of IT literacy will cause issues depending on the specific market. However, budgets will also be correspondingly tighter, particularly in relation to eHealth systems which have been developed to cater more to the Western market, and particularly as health budgets will be devoted more to basic health provision, medical supplies and manpower.

As mobile operators develop capability in the ICT space and begin to replicate the capabilities of an ICT infrastructure provider (such as IBM, EDS and Oracle). This may be acceptable in markets where the mobile operator has a natural incumbent advantage – but in other markets where this is not the case the key challenge then becomes one of differentiation.There are two parameters that provide mobile operators with a unique value proposition in the eHealth industry according to GSMA :

1. Leverage Global Integration capabilities for Supply Chain efficiencies

Traditionally, the mobile operator value chain consists of those core capabilities that enable it to acquire customers through its sales and distribution network, set them up on the network, identify and connect consumers on a network, create value added services in both voice and data, provide customer service and run sophisticated billing and tariff programs to optimise revenue per customer. In recent years, some mobile operators (particularly those with strong group operations) have developed specialised global business integration capabilities, ranging from cloud computing, portal technologies, payment mechanisms, Machine-to-Machine (M2M) platforms and solutions, and systems integration. These are the capabilities that allow the mobile operator to create solutions connecting the healthcare providers with the patient, as well as with other healthcare players in the industry, providing the industry level integration.

2. Shifting Demand with Integrated Participative Healthcare

In developing countries, the chronic shortage of healthcare professionals as well the prohibitive cost of building healthcare facilities in rural areas, are also indicators of a need for healthcare to develop beyond its traditional hospital-bound model. This can be done either by keeping patients from entering the system through effective prevention and wellness, or by managing patients consistently after they exit the system through medication adherence, effective monitoring and post-discharge management, particularly in the case of chronic diseases.

One key opportunity discussed is the ability to take mobile operators' core capabilities and apply them inventively to solve healthcare problems both large and small. Telefonica talks of a "lift and shift" economies of scale methodology whereby there is a rule about being able to re-use a significant percentage of the investment made into any particular technology in another region.". AT&T speaks of "utilising all of AT&T's core assets to apply to healthcare", and also of the importance of "scaling up" applications in both simple and complex settings, working from small to large scale.

Contingent upon the mobile operator's global business integration capabilities is its ability to offer health specific solutions e.g. Cloud-based PACS or records hosting, remote monitoring solutions to manage chronic diseases, or sophisticated tele-medicine capabilities incorporating collaboration technologies with remote diagnostic equipment. These will enable the mobile operator to differentiate its offering while leveraging on its core ICT capabilities. Depending on the mobile operator's business strategy, such capabilities can be developed either through internal development, partnering or acquisition. This strategy however requires greater internal investment in order to develop specialised health expertise, as well as to select suitable business partners with which there is mutual benefit.

Collaboration is an example of a capability that can apply both within the enterprise, between different healthcare institutions, as well as between the institution and the patient. At the simplest level, this can be web-enabled audio and video-conferencing solutions between physicians, to sophisticated solutions which incorporate both videoconference as well as peripherals which allow specialists to conduct diagnostic assessments to patients in remote areas. Cisco, for example, partners with both Telefonica and AT&T to provide such solutions. In Telefonica's case, the Health Presence product was introduced in order to alleviate the travel needs for patients in the Balearic Islands who previously had to travel to Mallorca for diagnosis.

Orange set up their Orange Healthcare business unit in 2007, talks of "joining up healthcare", echoing the role of mobile operators in connecting both the healthcare enterprise and the patient, as well as helping to remove the boundaries between the increasing number of players in the healthcare industry. They have organised themselves into services supporting the patient (in terms of wellness and prevention), to services supporting the healthcare professional and hospital operations, to services connecting the two in order to better manage the number one problem driving healthcare costs globally, that of chronic disease management.

Telefónica, which set up their global eHealth unit in 2009, speak of their unit as being a "standard bearer for health products and services", with three focus areas in Remote Patient Management (RPM), Telecare and Health IT. Grounded in the belief that healthcare both has a local context, as well as a need to cross-pollinate ideas and scale across geographies, their organisational strategy was to have separate individuals which have both a functional responsibility, as well as a regional/country focus

While the mobile operator industry relies a lot on partnerships to get into the health industry and expertise, the more established players do recognise the need to create their own in-house health expertise. When AT&T started their health business unit, for example, it hired 60% of their staff directly from the healthcare industry. Telefónica decided to build their team combining staff coming from the healthcare industry with experienced employees with deep knowledge of ICT as well as of the functioning and capabilities of a telco operator. Orange has 100 R&D professionals in healthcare in 4 Skill Centres globally in order to develop expertise tailored to healthcare systems in their global footprint. Telefonica centralises their technical resources in a Living Lab in Granada which houses all their healthcare application development expertise.

Cloud-based eHealth systems are gaining traction, as the number of locations of healthcare practice increase, along with the high cost of deployment and maintenance for traditional client-server models. These are particularly attractive to smaller clients with less complex needs, or clients with budget constraints. However, there are still residual concerns with the security and reliability of using a fully cloud-based solution for mission-critical health applications. Security and reliability can often be obtained through more "private" clouds which offer dedicated resources and guaranteed access, which unfortunately reduces the core cost advantage which cloud systems are procured for.Use case for the cloud is in the hosting of health records or acting as a connector between different enterprise level EMR systems. AT&T's Healthcare Community Online product is a health information exchange based on cloud technologies, supporting collaborative care through secure messaging, access to multiple applications through one portal, integrating with the American Medical Association's own portal.

Be sure of one thing " the mobile operator's evolving participation in eHealth will ultimately depend on the extent to which there is mutual value creation between the two sectors. For the mobile operator, there is the promise of a new corporate segment, a means to better utilise and monetise their enterprise ICT capabilities, an opportunity to extend their brand in the healthcare industry, and provide more subscriber value in their own mHealth services as they better integrate into the enterprise healthcare sector. SMART, a mobile operator based in the Philippines, recently rolled out a lightweight eHealth system in two major cities, with these constraints in mind.

A mobile operator based in South Korea, has also recently developed a hospital information system that it plans to roll out across various sites in its home country, and currently considering expansion to China.For the healthcare sector, the mobile operator's involvement represents an opportunity to reap the large economies of scale of using mobile operators' significant IT investment, and partnering with an ICT player who is best placed in helping it extend its reach to the patient.

Privacy and the integrity of personal and corporate data is becoming one of the most significant areas of the digital world, driven by trends such as BYOD increasing the use of cloud computing in the corporate space, or the continuous use of social media in the consumer space. As online fraud, e-crime and many emergent threats move into the tablet and smartphone space, a solid commitment to offering products and services to protect customers and their digital lives becomes paramount for any self respecting Telco.

Welcome to the dire need for Identity Management (IDM). In a nutshell, IDM is an integrated system of business processes, policies, and technology that lets an organization control user access to online applications or services while protecting the user's privacy and the organization's resources.Telcos that undertake strong IDM service programs can capitalize on their existing assets to create additional revenue streams with subscriber identity data.

In 2012, the Edelman Trust Barometer indicated that distrust amongst consumers was growing; compared to just a year ago, twice as many countries surveyed are now skeptical and fewer countries are neutral. In fact, a privacy survey of 9,200 interviewees in 14 countries around the globe conducted by the research company Psychonomics reveals that customers are increasingly concerned about the use and misuse of their data. 82% of respondents see privacy as an important topic; 76% are concerned about privacy violations; and 45% of subscribers feel they lack control over their personal data. Only 9% of respondents don't care about data protection in general, whereas 80% of respondents claim to be selective about sharing their personal data and why. A further 73% are afraid of their personal data being sold to third parties. In a digital age characterized by complex¬ity and concerns about privacy and security, customers are increasingly drawn to the idea of a Trusted Identity Provider.

However, the major challenge that operators face in using their data to generate revenue, is that customer data is still siloed on legacy systems, due to their past merger and acquisition activity. Therefore, operator IT departments would need to invest in data integration, MDM (Mobile Device Management) , data quality tools and the necessary in-house expertise to use these tools effectively. If Telcos could leverage this subscriber data in real-time they can not only improve their service offerings, but can also create completely new revenue streams and support third parties in numerous industry sectors in improving their service offerings to their customers.

Before Telcos can become true and trusted identity providers, they need more than just authentication, authorization, and accounting (AAA) servers inside their networks; they also need directory servers, tools for managing subscriber access, and the federation software itself. Thats big investments but if the Tecos don't do it the Web 2.0 players will do and capitalise on this huge IDM opportunity. Most executives compare the cash flows from innovation against the default scenario of doing nothing, assuming—incorrectly— that the present health of the company will persist indefinitely if the investment is not made. For a better assessment of an innovation's value, the comparison should be between its projected discounted cash flow and the more likely scenario of a decline in performance in the absence of innovation investment.

Some Telcos may not really know yet how they will make money as an identity provider, but they are certain that IDM will help deliver services that can reduce customer churn. Fortunately proactive operators do see this opportunity and are acting on it. France Télécom – Orange is among a handful of European and Asian network operators that are implementing IDM technology and learning to become identity providers – companies that can, in effect, vouch for a user's identity in transactions with Web-based service providers. In Argentina, operator MoviStar, is offering a single sign-on service. With one single password, the user can access to online services such as Facebook, Twitter and Flickr moving to "zero sign on", with all a customer's identities stored on the secure element of the SIM card.

Telfonica has segmented its Security market under Information Security , Electronic security and Mobile Security mapped across Consumer / SME and Enterprise / Gov sectors and sales estimated at $ 300 million. Their Enterprise security solutions include Web security gateway (secure internet surfing) , Clean e-mail (mail security service) , Anti-DDoS(preventing Distributed Denial of Service attacks) , Anti-phishing and fraud prevention and Managed Security services. The Device Management Service is available in a tiered 'pay as you use' structure incorporating:

• Device inventory: know what you've got and where it is
• Application management: update, remove and review software across devices
• Capability control: enable or disable functions or capabilities on devices
• Advanced security: keep your data safe with encryption, password management and firewall

The good news for operators is that it is not too late to join the race.A Telco's success in generating revenues as Identity Brokers depends on upon a close alignement with Enterprise verticals such as Healthcare , Finance and Goverment. With their many assets, operators are in a very good position to take a leading position in the Internet value chain. The operator can also act as " Trusted Identity Guardian " for end users and as part of a larger commercial ecosystem.

MPLS VPN is a family of methods for harnessing the power of Multiprotocol Label Switching (MPLS) to create Virtual Private Networks (VPNs). MPLS is well suited to the task as it provides traffic isolation and differentiation without substantial overhead. Healthy growth in the market is expected as organizations increasingly look to IP MPLS VPN services to control WAN costs in the face of rising bandwidth needs, to Ethernet for the lowest cost-per-bit, and to managed layer 2 and layer 3 services, where the expertise, knowledge, and tools of service providers can help stabilize WAN costs and prioritize critical applications while increasing capacities

Global Ethernet and MPLS IP VPN service revenue grew a combined 13% in 2011 to just over $50 billion, fueled by surging data traffic, cloud services, and cost-cutting initiatives. Asia Pacific is expected to overtake EMEA as the leading region for IP MPLS VPN services in 2013 and already leads in Ethernet services; Asia will remain the leader for the combined IP MPLS VPN and Ethernet services market going forward, led by China and India. IPsec VPN services accounted for the remaining share of 32.5%. In Middle East and Africa, MPLS VPN services accounted for largest share (67.5%) within IP VPN market in 2012.

Concurrently carrier Ethernet Exchanges are a significant new development that facilitate Ethernet connections and accelerate the move to Ethernet transport and services. In these exchanges service providers pay small fees to a Carrier Ethernet Exchange to make it easy for them to locate, buy, and provision Ethernet connections from each other. This in turn jumpstarts more Ethernet services and more of the IP VPN services that ride on Ethernet transport. The net effect of these new Ethernet exchanges, combined with fast-rising mobile backhaul connections, is a quickening of the Ethernet and IP VPN services markets

To capitalise on this growth opportunity BT Global Services invested in the rollout of new IP MPLS infrastructure, services, and expertise into Turkey, the Middle East, and Africa. On the face of it, the Middle East and Africa is a sprawling super-region, with as diverse a range of markets as could be found on the rest of the planet. The startups and traders of the narrow streets of Nairobi are a world away from the huge family-owned conglomerates of Turkey or the state-owned petrochemical giants of Saudi Arabia. Yet these enterprises have important things in common: they are part of the same global supply chains in manufacturing, transport, and logistics, and they need high-security broadband access and hosted applications. That makes the MEA a perfect target for network operators with the capability to combine their own UC platforms with those of vendors, and to provide both remote access and IT support

MPLS VPNs offer the ability to prioritize applications such as VoIP by class of service (CoS), create and improve disaster recovery infrastructures, utilize a fully meshed infrastructure that replaces outdated hub and spoke architecture, and reduce complexity to simplify network management in an increasingly complex landscape. The desire to move toward a more cost-effective network that supports voice, video and data is among the primary drivers behind the move to MPLS VPN services, with a number of other technological and financial drivers drawing enterprises in that direction as well.

All businesses need communications networks which accurately mirror their own changing data flows and their own transactions with customers. Enterprise clients need to connect site to site and function to function: the network infrastructure must enable multiple departments including sales, marketing, production, logistics and distribution to work together as a single business machine. And they want to do so in a way which minimizes cost and maximizes reliability.Among these are:

1. Class of Service (CoS): provides ability to prioritize applications, such as VoIP

2. Automatic redundancy/disaster recovery: create and improve disaster recovery infrastructures

3. Fully meshed infrastructure: replace outdated hub-and-spoke architectures

4. Reduced complexity: one network platform supports all application traffic— including VoIP and data applications

IP VPNs are typically utilized by organizations operating out of multiple office locations that require a secure, flexible and cost effective means for their employees to communicate and share information across a central computer network. MPLS networks form a proper foundation for a number of business critical applications including: VoIP phone service + Centralized merchant transactions + Remote application access (Citrix)

+ Remote user access + File transfer/sharing + Video delivery + Secure access to internal software applications + Outsourced network management + Compliance requirements + Central data storage & backup etc. Any business operating across multiple locations can benefit from a MPLS wide area network.

With the emergence of VoIP phone service, private wide area networks are serving a very important role in ensuring call quality, performance and security across the enterprise in several verticals such as : • Retail Businesses • Restaurant Chains • Hospitals and Clinics • Doctor and Dentist Offices • Call Centers w/ Remote Staff • International Corporations • Financial Institutions • Government Entities • Hotels • Franchises

Internet Protocol/Multi-Protocol Label Switching (IP/MPLS) has grown to become a foundation for many mobile, fixed, and converged networks. In mobile networks, IP/MPLS consolidates disparate transport networks for different radio technologies, reduces operating expenditures (OPEX), and converges networks on a resilient and reliable infrastructure, while supporting evolution to Long Term Evolution (LTE) and Fourth-Generation Mobile Network (4G) technologies.

BT seems to be taking the right approach by promising Enterprises with an extended infrastructure based on subsea cables in the region, new fiber connections into South Africa, and greater domestic connectivity. New MPLS nodes in Oman will extend the global reach of MPLS, while new NNIs (network-to-network interface) will take services out of South Africa into 12 other countries. In addition Ethernet managed services will be offered in four countries and a center of excellence for satellite established in Turkey. MPLS-based IP VPN and IPsec VPNs will be standard, accelerating Ethernet into 20 cities.

This will be supported by portfolio expansion, with 10 launches in each center combining regional and global products with some local integration. A key example of this is global inbound services. BT managed security services will be available in all countries.Being able to deliver on-net access in close proximity to a customer's local offices, subsidiaries and partners all contribute to an eff ective international IP/MPLS strategy and can result in a carrier being more competitive with regard to faster local provisioning and troubleshooting, and more cost-eff ective circuit fees.

According to Current Analysis GNT the ability of an operator to provide a geographically comprehensive on-net IP/MPLS footprint that can support a customer's key sites is critical for customers when procuring global network services.Unlocking the various revenue streams out of a carrier grade IP MPLS network requires a highly tuned understanding of enterprise needs and PoP location. Being able to deliver on-net access in close proximity to a customer's local offices, subsidiaries and partners all contribute to an effective international IP/MPLS strategy and can result in a carrier being more competitive with regard to faster local provisioning and troubleshooting, and more cost-effective circuit fees.

Carriers with strong in-region MPLS PoP ownership should certainly present to customers the advantages of working with a provider that can more effectively monitor and manage its own network for better performance and rapid response and troubleshooting. Ongoing efforts to expand networks and invest in PoPs in new regions, Ethernet VPLS expansions, new Ethernet and DSL capabilities and new NNI agreements will no doubt bolster the image and credibility of the MPLS provider.Naturally, having the highest MPLS PoP count is not enough and the service wrap, pricing and SLAs are equally important; carriers must recall that Vanco, as a VNO, managed to win customers fairly effectively until it was bought by Reliance Globalcom in 2008.

The main point here is that operators must be competitive on price and provisioning times and reassure customers that they are skilled in managing and selecting local network partners and that the partners they have chosen are reliable, financially stable and can deliver the target SLAs and required resiliency for a customer's sites.

Shared data is a success story for savvy mobile operators. In today's multidevice ownership market having a pool of data that different customers, with many mobile devices, can share is proving a win-win for operators and customers. In order to work correctly and deliver the desired results there are some basic building blocks that shared data plans need to be incorporated into the Telco OSS/BSS .

These technical foundations include   Real-Time Usage Tracking and Balance Management ; Shared Allowance Profile Management; On Device Plan Self Management ; Adding parental Controls and Offer Management via a catalogue. As most of us already know that the central importance of real-time capabilities in OSS/BSS permeates all aspects of operational, network and business management. Since many new services demand real-time support, operators must either transition to real-time OSS/BSS capabilities or forego these revenues. According to Current Analysis Operators will be "using virtualization to drive innovative service creation, especially the creation of services and apps that require time to market intervals of only days, even hours.

There is a wide range of applications of shared data. These include multidevice plans, family plans, group plans and business plans. There are two foundations of success for shared plans. First is giving the customer control in setting the plan and second is real-time balance management. With many different users and devices using the same data pool, it's vital that balances are managed in real-time, and that customers are fully in control of their usage and have real-time visibility of their charges. A look at the Q2 2014 results of some of the main innovators of shared data underlines the benefits that operators are realizing.

AT&T's Mobile Share shared data plans, now represent more than 41 million connections, with the number of Mobile Share accounts more than tripling year-over-year to reach 14.6 million, with an average of about three devices per account. 49% of Mobile Share accounts had 10 GB or larger data plans, up from 25% in Q2 2013. Mobile Share has helped drive year on year increase of 20% in wireless data billings. Year on year Verizon has increased revenue per account by 4.7% and growing the percentage of accounts on More Everything plans from 36% to 50%.AT&T's plans cover up to 10 devices per Mobile Share Plan and range from 300MB to 50GB of data to share. Plans included unlimited voice and SMS, and AT&T also offers 50GB of free cloud storage with AT&T Locker, which is marketed as a secure and safe place for customers to store their photos. When it comes to devices to add as well as smartphones, gaming devices, tablets etc.

Operators want customers to use tablets on their networks. Tablets are driving subscriber growth and operators are rolling out innovative offers to get customers buying and using cellular enabled tablets. As an example of contributions tablets can make to an operator's results, of the 1.4M retail net customers Verizon added in Q2 2014, 304,000 were postpaid phone net additions and the remaining 1.15 million were postpaid tablet subscribers. Mobility is a key driver for mobile data connectivity on tablets. This is particularly the case in emerging markets, where a higher proportion of tablet users than in more-mature economies report using their devices outside the home and on the move.

While the global tablet market is stabilising (IDC forecast 2014 worldwide tablet shipments of 233.1M units: a 6.5% year on year growth rate, after several years of double and treble digit growth), the number of cellular enabled tablets is on the rise. In Asia-Pacific, according to IDC, 25% of total tablet units shipped in the region have built-in option of voice calling over cellular networks (a 60% year on year growth). Shipments of tablet PCs to South Africa increased 107.1% year on year in the final quarter of 2013 to total 513,000 units .

In August 2014 T-Mobile also launched a tablet promotion for its customers on their Simple Choice plan. T-Mobile are matching the amount of data of a customer's smartphone to their tablet for $10 a month (up to a limit of 5GB). This level of pricing shows how eager operators are to get customers using tablets on their networks. T-Mobile's messaging pushes ' no overage charges' (they throttle speeds) in a drive to increase cellular tablet usage.Consumers increasingly watch TV and video on tablets. This will not be lost on operators who are looking to offer entertainment services—such as LTE broadcast and roll out TV / video partnerships. Getting tablet subscribers on board now may help ease the launch of these services as operators look to offer entertainment bundles to existing tablet subscribers.

One of the pioneers of shared data plans, Bell offers 'Family Shareable' and 'Personal Shareable' options. Customers can connect up to 10 devices or family members to the share plan, and also offer Mobile TV as a plan add on. The Mobile TV add on enables customers to watch over 30 live and 14 on demand TV channels for $5/ month per device for 10 hours. Bell's $50 and $60 share plan includes the Mobile TV add-on free for 3 months.

Telstra Australia launched their "shared" data plans for consumer plans in 2013. The plans are pretty simple once you get your head around it. Basically, it's available as a bolt-on option with Telstra's Every Day Connect consumer-level plans. It's not included for free, however: you'll be paying for the privilege. Every Day Connect plans come with one SIM card by default, but you can have up to three SIM cards connected to the one plan for data sharing with your SIM-enabled tablet. For example, say you have a phone with Telstra and an additional SIM-enabled tablet. The $60 Every Day Connect Plan gets you $600 of calls, unlimited text and MMS plus 1GB of data. To activate data sharing, you'll pay $10 for the Every Day Connect Data Share Plan, then an additional $10 for the additional SIM card for a final cost of $80 per month.

Sprint's Family Share Pack allows customers to connect up to ten devices or family members to a shared data bundle. Data bundles start at 600 MB and run all the way to 60 GB along with free unlimited voice and text. As well as their smartphones, customers have the option of connecting tablets ($10) and other mobile broadband devices ($20) to the plan. Where this offer is particularly interesting is in relation to new customer retention, Sprint will pay termination charges of up to $350 for customers porting over to this plan from another operator and also waive any access fees. This new plans has a limited running time, finishing at the start of 2016, which suggests Sprint are looking to entice new customers to sign up as well as make their current customer base "stickier".

In the UK, Vodafone is offering their own shared plan called "Red+". The plan aims to allow up to nine separate SIMs to connect to one "group leader plan". The group leader signs up to a data plan with an allowance of either 2GB, 4GB, 7GB, 10GB or 13GB and then defines how much of this allowance each member should receive. The plan is aimed primarily at families where the group leader would be one of the parents. Red+ plans allow the leader to both cap or expand the usage of each member, which is perfect for a parent whose teenager is a "data hog", and also offers free calls and texts to all members within the group. The really interesting part of this offer though is the fact that not only are there data notification alerts at 80 and 100%, but also you can't go over your allowance, unless you add extra data, meaning the subscriber is always in control.

UK 4G operator EE promotes flexibility in their 4GEE share plans, by enabling customers to add people to their plans at any time they want. The message is that if a friend (potential share plan member) is in the middle of a contract (presumably with a competitor) then the customer can add them when that contract finishes. This is a good example of using share plans to attract new customers. Starting at 250MB for $15 going up to 100GB for $750 Verizon's More Everything data share plan also provides unlimited talk and text offers. It also is offering a range of add-ons as standard – e.g. the plans come with 25 GB cloud storage and American Football (NFL) app offering NFL content and live games, an educational tools app, as well as the ability for customers to use their smartphones as a personal hotspot to get wi-fi enabled devices online.

There is no doubt that in the face of declining revenue from voice and messaging services, operators look for ways to monetise on data services and for pricing models that encourage customers to stay with their provider. Offering customers the option to sign up several devices with one subscription is not only attracting subscribers to sign up more devices (for example, by acquiring a connected tablet rather than a Wi-Fi only tablet), but also provides customers with an incentive to stay with one provider, if a single subscription provides cost savings, better matches their data consumption behaviour and facilitates the billing process.

In developed and developing countries alike, market saturation in telecoms is limiting customer acquisitions and value added services have not been able to generate the same revenue as voice services.Cloud offers a unique opportunity to service providers that want to offer value added services like voice, video and collaboration on cloud platforms, but success will come only with simplicity and a recognition that the economics of the cloud are very different than traditional telco models. You maye be surprised to learn that cloud models have utilization patterns like airlines which means : they are capital intensive, time and context sensitive and as such supply and demand differentiation becomes critical to maximizing yields.

Telcos today have around 5% of the public cloud market, according to analysts' estimates. They could potentially increase their market share by going beyond the provision of connectivity to provide additional services, such as authentication, billing, systems integration and even professional IT services. While most telcos can't match the IT expertise of IBM or HP, they have some advantages, such as long-standing relationships with both large and small businesses, well-known brand names and extensive customer care facilities. Technologies are considered to have become "mainstream" once they have achieved 25% penetration. As cloud follows this same trajectory, with a slew of telcos, cable operators, data centre specialists and colocation providers entering the market, significant consolidation will be inevitable, since cloud economics are inextricably linked to scale.

Cloud Computing attracts investments and overseas businesses and provides a significant boost to e-government initiatives. A Cloud Readiness Index is a good idea since it can track Africa's cloud adoption progress . By mapping the conditions and criteria required for successful implementation and uptake, one can identify potential bottlenecks that could slow cloud computing adoption. This "Cloud Readiness Index" would analyze key criteria that impact the deployment and use of cloud computing technology across different countries/cities : such criteria might include : Regulatory conditions , Data protection policy , Broadband quality, Power grid quality, Internet filtering , Government prioritization and ICT policy etc.

Without a doubt countries with the most insightful, transparent and fair regulatory environments supported by a the highest political echelons will be the most successful in capitalizing on this new opportunity. Using this criterion only 2 African countries seem ready for full blown Cloud Computing : Rwanda where the President is leading the broadband charge followed by Kenya which is emulating Rwanda : Fttx + Open Access 4 G and Govt sponsorship. With ICT for Development representing an area of high interest for the International Telecommunications Union, the World Bank, and other development agencies, it is inevitable that the services of cloud computing would be applied towards various areas of socio economic development. These area include e-education, e-health, e-commerce, e-governance, e-environment, and telecommuting.

African terrestrial and wireless broadband networks are ramping up catalysed by the arrival of several submarine cables at various landing points along the continental coastline. What may slow cloud uptake is lack of 4G Digital Dividend Spectrum and delays in terrestrial fibre without which Mobile Cloud Computing will remain a sluggish experience. Unreliable networks undermine the entire cloud concept and unfortunately most networks in Africa are battling with infrastructure bottlenecks and capex constraints.

Mobile operators can build their own dedicated, private cloud or they can outsource it to a third-party to host. Operators around the world are giving strong consideration to outsourcing due to the economies of scale that cloud service providers offer and the capability to shift CAPEX to OPEX to better manage costs. As providers of cloud services, telecom operators can manage connectivity, deliver cloud capabilities, and leverage network assets to enhance cloud offerings. Given their core competency, managing cloud connectivity appears to be the most natural value-adding activity.

Identity management and security also came through as strong themes and there is a natural role for telcos to play in the cloud Telcos already have a trusted billing relationship and hold personal customer information. Extending this capability to offer pre-population of forms, acting as an authentication broker on behalf of other services and integrating information about location and context through APIs would represent additional business and revenue generating opportunities. Another opportunity is driving productivity and efficiency gains for Enterprises, together with improved customer service and increased revenues : by allowing them to incorporate CSPs' communications and context-based capabilities directly into business applications such as Field Service Management.

Recent forecasts suggest that there will be up to 50 billion mobile connected "machines" over the coming years, including appliances, smart meters, security systems, healthcare devices and many others – all of which can benefit from network capabilities accessed on-demand from the mobile cloud. A global pioneer in M2M, Norway's Telenor Connexion is offering its M2M services through a dedicated platform that enables Telenor Connexion to focus on delivering highly responsive market offerings and developing differentiating value-adds, such as customisation services closely tied with connectivity. It serves M2M customers in automotive, fleet management, security, utilities and healthcare.

With a culture of aiming for five nines (99.999% uptime) reliability, telcos are well-suited to the delivery of cloud services dependent on continual connectivity. By leveraging their network assets, operators add value by exploiting user attributes such as profiles and activities, making cloud services relevant and meaningful to users and providing the linkage between the upstream and downstream components of two-sided business models. Telcos clearly have a pivotal role in the cloud value chain and Verizon Communications, Deutsche Telekom , SingTel , Etisalat and other telcos are moving aggressively into this market.

For example Telefónica was facing growing pressures on its revenues and profits. Meanwhile, companies in the critical field of small and medium-sized enterprises (SME) were requesting new services to strengthen their business capabilities.Telefónica gradually settled on software as a service (SaaS) as the potentially optimum way to meet the needs of these customers quickly while keeping costs to the minimum. The two key cloud platform technologies to satisfy Telefónica' s demands regarding its SaaS solution were "Aggregation Skills" and "Multi-Tenancy ".

The aggregation skill includes not only technology but business processors to aggregate and bring applications to the platform. These make it possible to deploy new applications very quickly and even globally if required thereby enabling the Telco to achieve benefits from the economies of scale . Multi-tenancy enables telco systems to accept upstream customers as operator-like entities, which could inject their own business rules into the system, use its development APIs, and run their product management independently.

Sometimes becoming a cloud service brokerage and bundling those services with their existing traditional telecom services confers certain advantages. One of the main benefits is service velocity, as XaaS services (Software-as-a-Service, Infrastructure-as-a-Service, etc.) can be on-boarded for pricing, fulfillment and billing in weeks, as opposed to the typical months-to-years timeframe. In addition, the costs to implement the services are dramatically reduced as well. That's what Deutche Telekom did with their Business Market place cloud for SME market in Germany.

Just as telecom operators are promoting cloud as a change agent for business, they too can benefit from its adoption. With operators seeking to transform themselves from their legacy environments and mindsets, adoption of cloud services can lead to efficiency gains, operational flexibility, and substantial cost savings. Perhaps that is where African Telcos should focus on to start with while the spectrum , fttx , Government support and other bottlenecks are sorted out.

Some key questions that must be answered while developing Telco Cloud strategies include : How can we best monetise the new service mix? How can the sort of margins available from voice services be realised for data? How can our business model evolve towards the sort of media retail models proposed by so many industry commentators? France Telecom Orange believe that in the initial phase Cloud services may not generate that much direct revenue but it reduces churn and that is good enough benefit to start with.

A ubiquitous mobile cloud will benefit the telecoms industry as a whole, by making it much more attractive for application service providers to create new services, or to enrich existing services, that use capabilities, information and intelligence provided by mobile and fixed telecoms operators. Eliminating fragmentation will result in a much larger addressable market of ASPs, resulting in increased service innovation, customer satisfaction, and new revenue sources for the industry as a whole, and consequently for individual operators.

The global entertainment market is huge and as such is obviously attractive to telcos looking to counter falling ARPUs. It accounts for a considerable share of disposable income and overall entertainment spending is much higher than that on telecoms. Online video makes up one-third of consumer internet traffic today and is forecast to grow more than ten times by 2013 to account for over 90% of consumer traffic overall.New consumer content consumption models that shift both time and place challenge operators to rethink their service models. Smartphones, tablets and Internet-connected TVs are carrying the feature set and expectations previously reserved for set-top boxes.

Believe it or not the single most important factor for success in OTT video is an attractive content library. However, these content rights are still cumbersome to acquire. This is due to the fact that OTT video rights form an entirely new category and that the value of such rights is yet to be determined. Consequently, content rights are negotiated on a piece-by-piece, geography-by-geography, business-model-by-business-model basis. Cooperation with content providers should be the primary choice in order to limit content procurement complexity and to achieve better economics of the service

So say hello to UltraViolet : No...not the sun's rays or the movie but an Outlier with Black Swan potential !! UltraViolet (UV) is a digital rights authentication and cloud-based licensing system that allows users of digital home entertainment content to stream and download purchased content to multiple platforms and devices. UltraViolet adheres to a "buy once, play anywhere" approach that allows users to store digital proof-of-purchases under one account to enable playback of content that is platform- and point-of-sale-agnostic. UltraViolet's basic proposition of 'buy once, play everywhere, forever, for the whole family' is a new and valuable one that overcomes many of the frustrations consumers have with online video content as it offers:

• A single point of access to content from multiple content owners

• The ability to buy once and view content on up to 12 devices

• The ability for up to 6 family members to view the same content

UltraViolet does not store files, and is not a "cloud storage" platform. Only the rights for purchased content are stored on the service. By creating a digital-rights locker rather than a digital media storage locker, UltraViolet bypasses the cost of storage and bandwidth used when the media is accessed and passes that cost on to various service providers. UltraViolet is deployed by the 74 + members of the Digital Entertainment Content Ecosystem consortium (DECE), which includes film studios, retailers, consumer electronics manufacturers, cable TV companies, ISPs, network hosting vendors, and other Internet systems and security vendors, with the notable exceptions of Disney and Apple.

DECE members developed a common file format (CFF) designed to play in all UltraViolet players and work with all DECE-approved DRMs. The format is based on existing standards from MPEG, SMPTE, and others, and was originally derived from the Microsoft

(PIFF) specification. The goal was to avoid the problem of different file formats for different players and to make it possible to copy files from player and player.

Put simply, UltraViolet is DVD for the Internet. Just as the DVD logo means that you can buy a DVD from any seller and expect it to play in any player with a DVD logo (DVD players, DVD PCs, DVD entertainment systems in automobiles, and so on), the UltraViolet logo means you can buy UltraViolet movies from any seller, keep track of your "online locker" or "virtual collection" of movies, and expect them to play on anything with the UltraViolet logo (PCs, tablets, smartphones, Blu-ray players, cable set-top boxes etc).

Telcos have a unique set of assets that enable them to become a true partner with content owners, not just another retail channel. These include: the ability to provide payments, multi-modal content delivery, support a user experience across three screens (TV, PC and mobile), provide interactive marketing and CRM data and support hands on customer care. At the same time, the increasing availability of enabling technologies, such as CDNs and adaptive bitrate streaming, means that even legacy infrastructure is no longer a deterrent to OTT video and that a consistent single-digit, megabit-per-second DSL connection is enough to support most video streaming activities. Ofcourse 4G and FTTx makes HD even more widely consumed.

It is worthwhile noting that the business case for entertainment products is not based on sales but instead on " foot traffic " and the overall size of the " shopping basket " just like in a Retail Mall. Many upstream services rely on the ability to secure consumer attention and sell this on to third parties in some form (a.k.a Telco 2 sided business model) .This is the basis of the advertising-based business models, including the one that dominates the Internet.

In the final analysis the profitability of an OTT video business depends on a large subscriber base (which is what Telcos have) and the highly efficient operations of a large scale technical platform (which is what Ultraviolet aspires to become). For Telcos that don't have an entertainment platform, UltraViolet offers an opportunity to join the party and use that infrastructure to access what is expected to be premium content which they can offer to customers through their own retail portals. For those that already have their own platforms consideration should be given to adding UltraViolet into the mix for what is lost in duplicating infrastructures could be gained with premium content. As for Telcos who chose to ignore UltraViolet and Digital Lockers may well find themselves deprived of premium content as some device manufacturer or broadcaster steps in to capitalise on this opportunity.

Telecoms operators in Western Europe could generate an additional €2 billion in profits from working with over-the-top (OTT) companies if they are willing to take an open platform approach to collaboration, according to recent study published by Northstream.

In adjacent business lines (Asymmetric business models) that rely on the telcos' network and IT service platforms, it makes sense to cooperate and place selective bets in order to grow and cultivate a distinctive value proposition. To what extent telcos succeed, however, will naturally depend on their own capabilities, their potential to become market leaders and whether the markets they serve follow local or global rules.

Regional telco groups such as Telefónica, Vodafone and Deutsche Telekom and national champions such as Telecom Italia, KPN and Swisscom are setting their sights on similar areas of growth: payment services, advertising, energy services, smart home services and healthcare services ("e-health"), to name but a few. To take the latter segment as an example, many patients may not yet be able to imagine having, say, their blood pressure measured by a mobile service provider. Yet Telecom Italia is using precisely this kind of remote monitoring system to save older patients the hassle of spending time in hospital and to save the national health system money.

E-health may not be a large market. But given that healthcare is heavily dependent on national legislation, it seems a sensible niche to occupy as global OTT players are unlikely to be interested in developing suitable offerings for each and every country. European telco groups might even have the chance to establish a European standard – although there is as yet no sign that telcos plan to prioritize such opportunities in collaboration with healthcare partners.

On the other hand, payment services and advertising business force national incumbents to line up against international competitors. Payment by mobile phone is regarded as a market that has so far been neglected but could be worth billions. Things are slowly beginning to move. Japan's NTT, for example, is already in possession of bank licenses in Europe. Deutsche Telekom, Vodafone and O2 have come together to form the "mpass" joint venture – a payment system for online purchases by mobile phone. Ofcourse OTT players have also joined the fray,

Why not the same windfall for Telcos in MEA if they were to adopt the platform approach of partnerships, and facilitates more innovative service offerings ??Bottom Line : You can't wish the OTT's away anymore than you can do the same with your mother in law !! ....So.... Adapt , Innovate , Collaborate or PERISH

--♠--

Telco Challenge # 6 : The Quest for Customer Loyalty

According to a recent Aspect Software study, almost 75 percent of consumers prefer to solve issues on their own—almost one out of three respondents noted they would rather talk to a toilet than a customer service representative. Mobile operators suffer from some of the worst levels of customer satisfaction in the world eventhough all the tools and platforms to offer superlative experience are available to them. What a disgrace !!!

The annual WDS Loyalty Audit has revealed that ONLY 35 percent of customers are highly satisfied with the mobile operator. More worrying for carriers is that a quarter of subscribers claim low satisfaction.The figures pose serious questions for operators, as feeling satisfied is intertwined with a customer's intent to repurchase. A unsatisfied customer is 8 times more likely to switch operators. Despite these disappointing figures, mobile operators still appear to be doing very little to understand their relationship with consumers or customer loyalty. This in turn means their loyalty programs and customer satisfaction schemes, vital to customer retention and solid business performance, remain outdated and fail to deliver.

Customer Experience Management is certainly one of the biggest current buzz words in the mobile arena. A lot has been said and written around CxM (sometimes called CEM) but still there is no unique industry consensus about what CEM actually includes. From managing the brand's perception across websites or street-shops to subscribers' actual appreciation about the services and applications they pay for, Customer Experience Management is claimed to be a part of every business process. A common misconception in the industry is that CxM is a replacement for CRM which simply is not correct.

As the industry moves into a growing market of digital services built on infrastructures that enable fast development and deployment of new services, the service portfolio itself is not sufficient to establish a lasting differential in the market place. Such a differential is quickly eroded by competitive service providers. Having tried to differentiate through technology and 'clever' pricing models and found the strategy to be short lived, service providers are realising that a more solid differentiation can be gained through managing the customer experience. This does not just mean delivering service that meets the customers' expectations but that all aspects of its business must support the concept of a superior customer experience.

Many countries have a penetration rate above 100%. In such a competitive market, churn has become the major concern of operators who have changed their priorities from customer acquisition to customer retention. Operators' financial reports show that, for a medium sized operator the average cost of retaining an existing subscriber amounts to tens of dollars per subscriber per year.Comparing this with "Cost Of Acquisition" for new subscribers (COA), often worth several hundreds of dollars per new subscriber, operators are inclined to pamper their existing subscribers, especially those generating higher Average Revenue Per User (ARPU). In this context it is key for operators to understand user expectations and adapt mobile data plans to their needs.

A successful transformation into the CxM world can only be achieved by building on top of good CRM processes and practices. CEM takes us a step closer to achieving improved customer satisfaction. Instead of asking the question, "This is what we are doing, how well are we doing?" which is a CRM approach, CEM asks, "What is important to you, and how well are we doing?. CxM is aimed at turning customers into fans by seeing the world through their own eyes. In the May 2012 issue of Telecom Buzz, published by MobileComm, the article "Customer Experience Management: The Next 'Buzzword'" declares the four pillars of telecom CEM ...no.... not nuclear physics or Astroflight.... but simply :

• Network experience (includes coverage, signal quality, speed)

• Commercial experience (includes billing, payment)

• Product experience (includes telecom products such as handsets, VAS)

• Service experience (includes after-sales service , customer queries)

In the social networking age you would already know the crucial role of social media and mobile marketing are the new building blocks when developing your future-proof CxM strategy. Increasingly Telcos are looking to looking to the social networking sites to provide valuable feedback on what the customer is experiencing. Twitter and Facebook provide rapid indicators on when things are going wrong. Systems that automatically monitor key social networking sites must be deployed to flag to the Service Management Centre when traffic increases. Often this is the first sign that a service is failing or a new service does not work the way that it should.

Delivering an effective Customer Experience Management requires a coordinated program across the entire organisation and is best achieved by adopting a maturity framework similar to the Capability Maturity Model Integration (CMMI) framework. CMMI is a proven process improvement approach whose goal is to help organizations improve their performance. The TM Forum is developing a maturity model for the implementation of CEM. This CEM model, as with CMMI, is a five stage model that guides the service provider on a journey to a fully implemented and controlled CEM environment.

The measurement of Customer Experience is based on measuring the extent to which the customer's needs are satisfied using customer/user centric measures such as: + Would advocate (e.g. churn and loyalty indicators) + Would recommend (e.g. Net Promoter Score) + Would Buy again + Product availability + Product usability.Having the right tools and OSS / BSS environments in place to support CEM is absolutely critical to achieving the end goal. As such establishing an early dialogue with tools suppliers (internal and external) has to be a priority in the early days of the program if for no other reason than the lead times for delivering and integrating the necessary solutions.

Bear in mind having CEM people without equipping them with supporting tools will lead to frustration and will feed the 'naysayers' with ammunition to criticize or undermine the CxM program. CEM is likely to introduce new working practices which may to some, seem unnecessary and a hindrance to rolling out new digital services quickly. Without strong governance the program will become disjointed with different parts of the organisation going their own way instead of a single 'joined up' approach to delivering a good customer experience.

Network management plays a major role in an improved CxM. By avoiding network congestion and poor performance, telecom operators improve the quality-of-service (QoS) level, which eventually can reduce churn to a great extent and increase customer satisfaction with an operator. However, an improved network QoS represents only one of the factors that influence the overall "customer experience" equation. Strategies for reducing churn need to take place at every step of the customer life cycle. If marketing communicates something that is not supported by product quality, network infrastructure, billing processes, or customer care/service teams, the relationship worsens until, ultimately, the customer moves away.

Recently Telefonica implemented a suite of CxM tools and platforms to improve the end-to-end customer experience across mobile data, mobile voice, IPTV, high-speed Internet, cable, satellite and voice services. The platforms and tools enable their customers to troubleshoot and manage their digital experiences through devices such as mobile phones, laptops and IP set-top boxes, via dedicated web portal and apps.Telefonica's approach towards managing the customer experience embraces a multitude of critical success factors including customer surveys, social media activity, contact centre stats and service specific data. Recognising that the CEM view is a complex but profitable undertaking if you get it right , forward looking Telcos such as Telefonica have developed an OSS/BSS environment that enables them to display disparate customer data in one single 'vital signs' view.From this single view Telefonica are able to calculate various Customer Satisfaction Index values which they can then use to drive their customer centric quality improvement programs.

For the telcos to remain competitive an overarching customer-experience strategy ultimately makes more business sense.With growing pools of data (both structured and unstructured), gaining individual customer insights and coming up with products and services that suit them would require a sophisticated level of business intelligence such as CxM, which would deliver performance analytics to all management levels. Telcos that continue to ignore the dire need of having an effective CEM strategy supported by appropriate tool sets might eventually find themselves thrown out of the race...so good riddance and the Telco industry will be all the better for it !!

Unless you live in a cave or on Mars, you should know that social is the way to go. Social media - blogging, online social networking, and micro-blogging - have become so pervasive that it is almost unthinkable for a business entity - at least those who want to remain relevant !! In telecom, social media have transformed not only business models but the very concept of customer service. Emerging markets have embraced social media with gusto. Both India and Brazil represent some of the most aggressive growth, where more than 90 percent of online survey respondents report having an account on a social networking site. The reasons for this social media explosion in the emerging markets can be attributed to the concentration of Generation Y and younger, the cultural emphasis on maintaining regular contact with friends and family, and the influx of mobile technologies.

The IBM Institute for Business Value surveyed more than 1,000 consumers worldwide to understand who is using social media, what sites they frequent and what drives them to engage with companies. What the results showed   may come as a surprise to those companies that assume consumers are seeking them out to feel connected to their brand. In fact, consumers are far more interested in obtaining tangible value, suggesting businesses may be confusing their own desire for customer intimacy with consumers' motivations for engaging. For companies that have been taking a "build it and they will come" approach to social media, these consumer findings are a wake-up call that much more needs to be done if they want to attract more than the most devoted brand advocates.

Telefónica Europe wanted to develop a European-wide social media strategy to align social media use across each of the company's business units and to help staff use it to enhance the whole customer experience. Telefónica Europe started by evaluating the company's current use of social media across all O2 branded businesses. Not only did this help to determine the strategic aims of each business unit, but, more importantly, it helped to highlight where the business was currently realizing value from social media, and the quick wins and opportunities for improving customer experience in the long run. Once the audit was complete, the next step was to develop a consistent strategy for the use of social media across multiple European markets in support of Telefónica's commercial brand, O2, and its Brussels-based Public Affairs department. This included developing three-year aims and objectives for Telefónica Europe's social business strategy; the whole process also took into account each local market's needs and conditions.

Salesforce.com was selected as the platform which allows a fast delivery of the solution in the cloud and also provides an integrated module for social media communication called "Chatter" lifting Web 2.0 Technology into the Cloud. Due to the scalability and multi-tenancy of the Cloud solution, further CRM applications can be easily integrated.Telefonica are developing, testing and embedding social media programmes across Europe in different departments and teams including Customer Experience, Brand, Communications and Customer Contact Centres. Education and training for 29,000 employees was implemented to embed the social media strategy and standardised processes across the business to ensure best practice and maximum return from social media.

As a leading Telco with over 23 million customers, O2 (subsidiary of Telefonica) decided to embrace social channels, as it is increasingly seeing a change in customers' buying and service expectations—with a growing preference to use the online channel. O2 started a business transformation journey to a multi-product, multi-channel company whilst continuing profitable growth. Evidence of this transformation can be seen in some of the services such as the new online shopping experience for small office/home office businesses. A Chatter app clears the line for employee communication O2's in-store support staff—called "gurus"— to collaborate on issues and help customers immediately, often while they are still standing at the counter.

In the Telco industry, loyal customers are the key to success. Nobody knows that better than Sprint, rated #1 for customer satisfaction and the third-largest telecommunications provider in the United States. When Sprint wanted to use new social media platforms to share information across groups and manage relationships with business customers more efficiently they consolidated customer information, automated processes, and built apps to make it easy to share data with retail stores. Information on business customers from multiple CRM systems was consolidated into customer profiles in the social media platform : a one-stop-shop for our sales teams for business processes and customer intelligence.

More than 6,000 employees now use a Cloud platform to track accounts, contacts, and opportunities for improved visibility and real-time analytics.Sales Reps can quickly identify the best opportunities driving new leads and reducing the time to close enterprise deals by 25%. And, a fully-integrated configure-price-quote system helps Sprint's sales teams quickly build complex pricing models using up-to-date account information.

Customer visibility is enhanced by a Chatter application, which helps reps share and information and collaborate on deals so they can address customer needs quickly and provide consistently great service. Using a custom Internet portal, retail staff can easily share business leads with sales, and no leads get dropped. In the future, retail personnel will be able to collaborate with customer service professionals to solve customer issues on the spot. Custom apps built on the social media platform automate processes including managing waitlists for hot new phones, scheduling corporate briefings, or tracking churn data, so reps can proactively reach out to customers in danger of defecting or counteract competitive promotions. Another app manages the discount approvals process and recaptures almost $70 million in unauthorized discounts each month.

The benefits are social media based CRM are deep provided it is implemented strategically. First, there is the social interaction itself, which can provide direct value to the business through revenue from social commerce and cost savings when used for customer care or research, for example. Plus, social networking enables rapid, viral distribution of offers and content that may reach beyond what could be done in traditional channels – all with endorsement from connections people trust. But that is just the beginning. Companies also can use social platforms to mine data for brand monitoring and valuable customer insights, which can spark innovations for improved services, products and customer experiences. In a constant cycle of listen-analyze-engage evolve, Telcos can optimize their social media programs to continually enhance their business. But :Telcos beware. Hell hath no fury like a customer scorned on social media !!

Shared data is a success story for savvy mobile operators. In today's multidevice ownership market having a pool of data that different customers, with many mobile devices, can share is proving a win-win for operators and customers. In order to work correctly and deliver the desired results there are some basic building blocks that shared data plans need to be incorporated into the Telco OSS/BSS .

These technical foundations include   Real-Time Usage Tracking and Balance Management ; Shared Allowance Profile Management; On Device Plan Self Management ; Adding parental Controls and Offer Management via a catalogue. As most of us already know that the central importance of real-time capabilities in OSS/BSS permeates all aspects of operational, network and business management. Since many new services demand real-time support, operators must either transition to real-time OSS/BSS capabilities or forego these revenues. According to Current Analysis Operators will be "using virtualization to drive innovative service creation, especially the creation of services and apps that require time to market intervals of only days, even hours.

There is a wide range of applications of shared data. These include multidevice plans, family plans, group plans and business plans. There are two foundations of success for shared plans. First is giving the customer control in setting the plan and second is real-time balance management. With many different users and devices using the same data pool, it's vital that balances are managed in real-time, and that customers are fully in control of their usage and have real-time visibility of their charges. A look at the Q2 2014 results of some of the main innovators of shared data underlines the benefits that operators are realizing.

ATT's Mobile Share shared data plans, now represent more than 41 million connections, with the number of Mobile Share accounts more than tripling year-over-year to reach 14.6 million, with an average of about three devices per account. 49% of Mobile Share accounts had 10 GB or larger data plans, up from 25% in Q2 2013. Mobile Share has helped drive year on year increase of 20% in wireless data billings. Year on year Verizon has increased revenue per account by 4.7% and growing the percentage of accounts on More Everything plans from 36% to 50%.AT&T's plans cover up to 10 devices per Mobile Share Plan and range from 300MB to 50GB of data to share. Plans included unlimited voice and SMS, and AT&T also offers 50GB of free cloud storage with ATT Locker, which is marketed as a secure and safe place for customers to store their photos. When it comes to devices to add as well as smartphones, gaming devices, tablets and so on.

Operators want customers to use tablets on their networks. Tablets are driving subscriber growth and operators are rolling out innovative offers to get customers buying and using cellular enabled tablets. As an example of contributions tablets can make to an operator's results, of the 1.4M retail net customers Verizon added in Q2 2014, 304,000 were postpaid phone net additions and the remaining 1.15 million were postpaid tablet subscribers. Mobility is a key driver for mobile data connectivity on tablets. This is particularly the case in emerging markets, where a higher proportion of tablet users than in more-mature economies report using their devices outside the home and on the move.

While the global tablet market is stabilising (IDC forecast 2014 worldwide tablet shipments of 233.1M units: a 6.5% year on year growth rate, after several years of double and treble digit growth), the number of cellular enabled tablets is on the rise. In Asia-Pacific, according to IDC, 25% of total tablet units shipped in the region have built-in option of voice calling over cellular networks (a 60% year on year growth). Shipments of tablet PCs to South Africa increased 107.1% year on year in the final quarter of 2013 to total 513,000 units .

In August 2014 T-Mobile also launched a tablet promotion for its customers on their Simple Choice plan. T-Mobile are matching the amount of data of a customer's smartphone to their tablet for $10 a month (up to a limit of 5GB). This level of pricing shows how eager operators are to get customers using tablets on their networks. T-Mobile's messaging pushes ' no overage charges' (they throttle speeds) in a drive to increase cellular tablet usage.Consumers increasingly watch TV and video on tablets. This will not be lost on operators who are looking to offer entertainment services—such as LTE broadcast and roll out TV / video partnerships. Getting tablet subscribers on board now may help ease the launch of these services as operators look to offer entertainment bundles to existing tablet subscribers.

One of the pioneers of shared data plans, Bell offers 'Family Shareable' and 'Personal Shareable' options. Customers can connect up to 10 devices or family members to the share plan, and also offer Mobile TV as a plan add on. The Mobile TV add on enables customers to watch over 30 live and 14 on demand TV channels for $5/ month per device for 10 hours. Bell's $50 and $60 share plan includes the Mobile TV add-on free for 3 months.

Telstra Australia launched their "shared" data plans for consumer plans in 2013. The plans are pretty simple once you get your head around it. Basically, it's available as a bolt-on option with Telstra's Every Day Connect consumer-level plans. It's not included for free, however: you'll be paying for the privilege. Every Day Connect plans come with one

SIM card by default, but you can have up to three SIM cards connected to the one plan for data sharing with your SIM-enabled tablet. For example, say you have a phone with Telstra and an additional SIM-enabled tablet. The $60 Every Day Connect Plan gets you $600 of calls, unlimited text and MMS plus 1GB of data. To activate data sharing, you'll pay $10 for the Every Day Connect Data Share Plan, then an additional $10 for the additional SIM card for a final cost of $80 per month.

Sprint's Family Share Pack allows customers to connect up to ten devices or family members to a shared data bundle. Data bundles start at 600 MB and run all the way to 60 GB along with free unlimited voice and text. As well as their smartphones, customers have the option of connecting tablets ($10) and other mobile broadband devices ($20) to the plan. Where this offer is particularly interesting is in relation to new customer retention, Sprint will pay termination charges of up to $350 for customers porting over to this plan from another operator and also waive any access fees. This new plans has a limited running time, finishing at the start of 2016, which suggests Sprint are looking to entice new customers to sign up as well as make their current customer base "stickier".

In the UK, Vodafone is offering their own shared plan called "Red+". The plan aims to allow up to nine separate SIMs to connect to one "group leader plan". The group leader signs up to a data plan with an allowance of either 2GB, 4GB, 7GB, 10GB or 13GB and then defines how much of this allowance each member should receive. The plan is aimed primarily at families where the group leader would be one of the parents. Red+ plans allow the leader to both cap or expand the usage of each member, which is perfect for a parent whose teenager is a "data hog", and also offers free calls and texts to all members within the group. The really interesting part of this offer though is the fact that not only are there data notification alerts at 80 and 100%, but also you can't go over your allowance, unless you add extra data, meaning the subscriber is always in control.

UK 4G operator EE promotes flexibility in their 4GEE share plans, by enabling customers to add people to their plans at any time they want. The message is that if a friend (potential share plan member) is in the middle of a contract (presumably with a competitor) then the customer can add them when that contract finishes. This is a good example of using share plans to attract new customers. Starting at 250MB for $15 going up to 100GB for $750 Verizon's More Everything data share plan also provides unlimited talk and text offers. It also is offering a range of add-ons as standard – e.g. the plans come with 25 GB cloud storage and American Football (NFL) app offering NFL content and live games, an educational tools app, as well as the ability for customers to use their smartphones as a personal hotspot to get wi-fi enabled devices online.

There is no doubt that in the face of declining revenue from voice and messaging services, operators look for ways to monetise on data services and for pricing models that encourage customers to stay with their provider. Offering customers the option to sign up several devices with one subscription is not only attracting subscribers to sign up more devices (for example, by acquiring a connected tablet rather than a Wi-Fi only tablet), but also provides customers with an incentive to stay with one provider, if a single subscription provides cost savings, better matches their data consumption behaviour and facilitates the billing process.

Big data has been a headline theme in the technology and mobile space for some time.Telcos all over the globe are seeing an unprecedented rise in volume, variety and velocity of information ("big data") due to next generation mobile network rollouts, increased use of smart phones and rise of social media. Telco operators have historically focused on managing the network with little visibility to the impact it has on the customer's experience. Which means the operator was forced to work with snapshots of network data in fragmented views or at a summary level in order to plan network capacity or provide information to customer care and marketing about customer transactions until now !!

" Offering free basic internet services can help mobile operators grow their businesses faster in emerging markets. And once users get a free taste of the internet they'll be more inclined to pay for mobile data.The Ebola crisis in West Africa is an example of a place where a crisis may have been exacerbated by the lack of good connectivity,"

 " (Mark Zuckerberg , CEO Facebook)

Big Data technologies, and in particular their analytics abilities, offer a multitude of benefits to telecom companies including improved subscriber experience, building and maintaining smarter networks, reducing churn, and generation of new revenue streams. Mind commerce, expects the Big Data driven telecom analytics market to grow at a CAGR of nearly 50% between 2014 and 2019. By the end of 2019, the market will eventually account for $5.4 Billion in annual revenue. Mobile commerce is one particular area where operators and service providers can potentially deliver tangible benefits from the application of big data analytics. The growth of m-commerce is creating large amounts of information on consumer behaviour and choices, which can be used to offer more personalised services and offers. SK Planet (Division of SK Telecom) have stated that "our Cash Bag m-commerce portal should generate $9.3 billion in revenues this year, and by using big data analysis we can provide customers with a much improved experience, and not based simply on offering the lowest price."

"While big data is great, there is a lot of it out there in silos, and each data [set] speaks its own language, you need to be able to solve issues around homogenising it, then you need to solve issues around analysing the data... what we need is a 'universal translator.We also need a better policy over who owns this data. We need to have joined up thinking about these open platforms," (Young Sohn, Samsung Electronics, president and CSO)

Big data analytics solutions enable service providers to analyze real-time location data over time for opt-in subscribers to understand subscribe lifestyle. Combining lifestyle and mobile profiles with subscriber usage and digital behavior allows service provider to create targeted offers for opt-in subscribers. With a majority of subscribers using smart phones to access data services as well as voice, mobile network operators are seeing explosive growth in traffic levels across their networks. In addition, the mobile network operator environment is fiercely competitive, with the ability to attract, retain and grow valuable subscribers being key. Increasingly, the provision of high quality customer care is an important component in the marketing mix and in retaining subscribers.

The growth of connected devices, particularly in areas such as the home or in the car, presents new opportunities but also challenges for operators and other ecosystem players. Users may be willing to share data with service providers but on the basis that the data is used securely. This year the GSM industry introduced a standardised mobile identity solution that aims to become the de facto single sign-on tool that consumers could rely on to authenticate themselves in both online and offline environments. This initiative is set to stimulate adoption of mobile services that rely on absolute confidentiality, such as healthcare, government and banking.

The announcement of 'Mobile Connect' has came alongside a number of other partnerships are aimed at accelerating the uptake of digital commerce services via mobile solutions. Mobile Connect is backed by 12 leading operators including Axiata Group Berhad, China Mobile, China Telecom, Etisalat, KDDI, Ooredoo, Orange, Tata Teleservices, Telefónica, Telenor, Telstra and VimpelCom, as well as key industry players such as Dailymotion, Deezer, Gemalto, Giesecke; Devrient, Morpho, Oberthur and VALID.

The Open Mobile Alliance is also developing standards and specifications for how devices and apps should share data with other connected devices, supporting the development of interoperable end-to-end mobile services. The move towards increasing standardisation is a key element in reassuring consumers and addressing concerns of privacy and security around personal data. The proliferation of smart phones presents new opportunities and challenges: consumers want the best deals for all purchases based on their real-time location while requiring the services provider to honor their privacy preferences and provide only relevant offers when requested/opted-in.

Given the highly secure capabilities of the SIM, mobile phones could become the perfect tool for future Digital Identity, not only for digital use cases but also for authentication in offline environments (national ID card, airport check-ins, etc). This makes Mobile Connect initiative ideal in a range of environments (both online and offline), yet this web-based authentication service does not necessarily need to be linked to a SIM to function.

Mobile Connect is a web-based authentication service runs on the OpenID Connect protocol, ensuring interoperability across mobile operators and service providers. The identification solution being developed will use the subscriber's mobile phone number or mobile user name and information contained in the secure SIM card, meaning that consumers will no longer need to create and manage multiple user names and passwords.

At a strategic level, the Mobile Connect service establishes the SIM card and the mobile medium as a frontline identity management service provider, allowing mobile operators to participate in the critically important e-commerce market. In the longer term, this type of standardised mobile identity solution will help operators to derive substantial revenues from their presence in fast- growing e-commerce markets, notably by extending the reach and presence of operators' brands, raising levels of awareness and ultimately, improving loyalty.

Big Data opens a vast array of applications and opportunities in multiple vertical sectors including, but not limited to, retail and hospitality, media, utilities, financial services, healthcare and pharmaceutical, telecommunications, government, homeland security, and the emerging industrial Internet vertical. In fact, according to Heavy Reading's Big Data ; Advanced Analytics in Telecom report, the industry will move from the $1.95 billion it spent on Big Data and analytics in 2013 to $9.83 billion in 2020. Thats really big business even if we are to take these projections with a pinch of salt. Ofcourse you will have to hire the right set of skills to make sense of and monetise the data deluge.

And while solutions are already present that will help you ride that wave of data, the question is: how can you profit from it? Here are some insights into how telecom companies can use the power of Big Data to their advantage.

1. Monetize it : Brands like Telefónica O2 are figuring out how to provide valuable analytical insights to customers to help them become more effective. Using mobile network data, telecoms provide valuable details into shopping habits.

2. Sell Subscriber Insight Data :   Other companies like ATT are looking at selling aggregated customer data to marketing and advertising firms.

3. Advertise Smarter on Mobile : Still other brands are using Big Data to strengthen their mobile ad campaigns. SingTel recently acquired mobile advertising platform Amobee, in an effort to help clients better reach their target audience and deliver more relevant offers.

4. Better Analyze Set Top Box Data : Using the data that today's sophisticated set top boxes offer may provide new revenue streams from targeted ad sales and more customized and personalized content services.

The Gurus at Strategy &; believe that many types of data are potentially available to operators — and certain sets of data might be combined to open up new business opportunities in areas such as campaign marketing and fraud prevention. Operators could generate more accurate and personalized offer recommendations for existing individual subscribers by combining internal structured data, such as how and where each subscriber uses his or her phone, with external unstructured or semi-structured data from social media platforms (for example, Facebook and Twitter).

Our customers trust us, so our responsibility of how we manage this private information will become higher. Operators will have to become involved in defining how this sort of data is used, while protecting the privacy of the individual."

(Kaoru Kato, Docomo's president and CEO)

By correlating internal location, usage, and account data with external sources such as credit reports, operators could significantly increase the detection of fraudulent activity such as looping or call forwarding on hacked PBXs (private branch exchanges), or fraud involving the swapping of SIM cards, and improve the overall accuracy and efficiency of their efforts to recognize patterns of fraudulent behavior.Imagine having the best of both worlds ? Having the tools to analyze the growing amount of data and content your business is generating, and also finding ways to make it profitable. If you are astute then this deluge of data isn't a threat; it's a serious opportunity to take your telecom business in a new, exciting, and yes, profitable direction !!

The desire to feel good is fundamentally different from the desire to do good. Feel good is when we give a few coins to the beggar at the traffic robot : do good is when we take the beggar off the street and give him education and a home. The same applies to Corporate CSR initiatives. It's a mindset that spells success or waste in CSR . Strategic CSR is a concept that is appreciated only by visionary and responsible minds and those maybe sorely lacking in the CxO suite in Africa : after all quantifying the dollar benefits of soft wishy washy programs is nebulous at best and a nuisance at worst. The ROI on CSR is a medium to long term initiative and how you create linkages to satisfy both commercial and social objectives takes effort and specialist skills.

In most cases Telco CSR activity in Africa is normally delegated to some middle manager with a stingy budget.The initiatves are normally designed for PR and / or to score brownie points with the Government ....and maybe to assuage a nagging conscience !! Handing a few branded T shirts and umbrellas feels good but it hardly constitutes an element in a strategic CSR plan...that was a corporate branding exercise duh !! When CSR is done with a DO GOOD mindset then initiatives that help reduce the environmental impact of the business and that of its customers; increasing connectivity to the markets that need it most; and in providing value added and community-enriching services such as healthcare and education.Without a doubt the Gulf Operators have taken a gigantic lead in the CSR / Green IT / Talent development game unlike their sluggish competitors in Africa.

UAE based Etisalat Group has outdone itself in " DO GOOD " CSR and Green IT arena. Their programs on issues such as environmental sustainability, alternative energy, clean technology, and social welfare are honed to a fine art : a symbiotic synergising circle between social development and business objectives .In 2012 the Etisalat Group's sustainability and social responsibility strategy was transformed as the company undertook several commitments to ensure that its efforts are better coordinated across all operations. This included pledging its support to the United Nations Global Compact.

One of the key highlight has been the success of Etisalat's Mobile Baby Programme in Africa - a mobile health initiative that is supporting pregnant women in rural areas. The programme, which was launched in partnership with Qualcomm, Great Connections and D-Tree International, has now saved hundreds of lives and is being offered in multiple countries including Tanzania and Nigeria. This initiative has captured the industry's imagination. The Etisalat Group along with other major operators, have joined hands to form a programme that will bring together the major African operators to work across borders to provide rudimentary healthcare services under the banner of the GSMA's Pan-African mHealth Initiative.

The folks at Etisalat have long been convinced that through encouraging diversity and equal opportunities, the company is more creative and innovative. This results in more satisfied customers, better corporate performance, and an environment that attracts the best talent.To ensure that Etisalat's business will be sustained with the best talent in the future, it launched an innovative Hi Potential (HiPo) programme including almost one hundred staff from across Etisalat's footprint.Etisalat's HiPo Programme provides these employees with access to valuable learning resources from organisations such as Harvard Business School and Duke University, as well as the opportunity to take work placements in other countries within Etisalat's footprint. In other cases and where the need is even greater (e.g. Afghanistan), a comprehensive succession development programme has been designed with key focus on preparing future leaders from within the Afghan national workforce. The programme is based on global best practice and to date has over 100 successors who are being developed for future leadership roles.

Etisalat's Go Green strategy forms the basis of its environmental management strategy and incorporates four stakeholder groups – staff, customers, government bodies and Non-Government Organisations (NGOs).Many of Etisalat's office buildings have smart technology installed throughout – ensuring that electricity is switched off at the end of day. Its buildings have also been designed by leading architects to reduce the heat effect and minimise the requirement for air conditioning.In 2010 Etisalat formalized its environmental policy and strategy to help focus efforts in reducing our internal consumption and promoting efficiency across our operations. This was launched under the theme of 'Go Green' and has so far involved:

• Paper collection and recycling • Photocopy cartridge recycling • Recycled paper replacing farmed paper for office use • Network printing and photocopying to reduce waste and manage consumption

Etisalat is pursuing tower sharing partnerships and deploying hybrid and alternative power solutions including wind and solar power, across its footprint. Significant projects are under development in Nigeria, Egypt and Afghanistan. These programmes will not only ensure that emissions are reduced and improve Etisalat's operating expenses, but will also provide greater resilience to the company's services.Their Energy Star Initiative cut the greenhouse emissions from buildings in the UAE. More than 3 million sq meters of facility space from 20 of the largest companies in the UAE are now being managed through the Energy Star project. It has also been calculated that over 3,000 tons of CO_2

emissions were eliminated in the first eight months of 2012, with participants seeing a 15-25% reduction in electricity consumption.

Not to be outdone by Etisalat's CSR leadership other Gulf Telcos are ramping up to provide information and education on more sophisticated business concepts. Emirates Integrated Telecommunications Company (known commonly as du) launched the QA platform Ejaba.me in 2013. The site provides entrepreneurs with answers from experts on topics such as management, legal services, funding, and intellectual property rights. Designed by French entrepreneur Joanna Truffaut, Ejaba.me is intended to give entrepreneurs access to affordable advice at a time in their development when funds are scarce. The site charges by the answer: US$20 for short answers and $30 for longer ones.

Other companies in the region are offering more hands-on training and mentorship to entrepreneurs in fields as varied as telecommunications and maritime operations. Dubai-based digital media company Intigral, a joint venture between Saudi Telecom Company (STC) and All Asia Networks (ASTRO), launched an accelerator named Afkar.me to support startups building digital products and services.

The biggest and most successful skills development initiative in Africa was the creation of the $ 2 million Motorola Cellular Training Institute and that was over a decade ago in South Africa. Yours truly conceived and implemented this as a strategic response to the growing technical skills crisis. Based in Johannesburg, the MCTI was the only one of its kind in Africa and one of only six in the world. Designed to train up to100 students at any given time, the facility was equipped with a cutting edge GSM /GPRS laboratory comprising state-of-the art equipment to simulate cellular networks on which students would hone their practical skills.The training curriculum was designed around a number of jobset descriptions based on extensive research into the skill needs of the cellular industry.

The robust portfolio of training courses fulfilled the needs of engineers and technicians in network operations, OMC systems, Network planning, BSS optimisation, BSS database and diagnostics, and cellular field engineers. A special certificate in wireless telephony has been developed in partnership with a local University to fast track electrical / electronic engineers, experienced radio engineers and computer engineers without cellular experience into competence as system design engineers, field engineers, optimisation experts for the cellular industry in sub Saharan Africa.

As the Telco network transforms into software platforms (Cloud , NFV , SDN etc) the MEA Telcos need to do more to accelerate the skilling software engineers by setting up Software Engineering Centres instead of relying on universities. What is urgently required is an innovative industry-university-government collaboration to prepare math and science graduates for advanced study in software engineering, telecommunications, and satellite communication and provide them with convenient advanced degree programs This will ramp up the development of software and telecommunications engineering human resources, and help accelerate the development of ancillary telecommunications and software engineering industries in MEA Region.

CSR (education , Green IT , poverty alleviation , job creation etc) should take centre stage in any self respecting Telco's business and sustainibility strategy. Corporate Sustainability is a business approach that creates long-term shareholder value by embracing opportunities and managing risks deriving from economic, environmental and social developments. Since corporate sustainability performance can now be financially quantified, they now have an investable corporate sustainability concept. Second, sustainability leaders are increasingly expected to show superior performance and favourable risk/return profiles. More and more, investors are convinced that sustainability is a catalyst for enlightened and disciplined management, and, thus, a critical success factor.

" Mobile connectivity hold huge potential for women – currently an untapped economic potential. The Internet is proving to be an effective catalyst in transforming gender opinions. Access to a communication network provides women with the flexibility of working both remotely and on their own terms, allowing them to build independent companies. The more we can work together to develop this offering, the greater opportunities we can build for the region's women as a whole" (Dr Nasser Marafih, Group CEO, Ooredoo)

---◆--

Telco Challange # 7 : The Quest for Softer networks

According to Deutche Telecom CEO communication networks are facing a lack of scalable and sustainable architecture to meet the challenges ahead in terms of data traffic increases, video uploads and downloads, and enhanced M2M communication. But employing software-defined networking (SDN) techniques could help mobile carriers overcome those hurdles and attract new data-centric revenue streams.In a nutshell, SDN delaminates the data and control planes of the network and NFV virtualizes the functional elements of the network—routers, switches, firewalls—and expresses these functions as programs that run on commercial off-the-shelf (COTS) IT hardware. While they are distinct technologies, the two work together in concert to turn the network into an infinitely programmable dynamic mesh, versus a hardware-based static map. Where SDN is the network admin gone virtual; NFV is the gear gone virtual.

Today's mobile networks are limited and built upon a best-effort design, but that means they have latency issues and cannot dedicate high bandwidth to a particular user on the fly. Network virtualisation highlights the transformational path that operators are willing to take to counter the stress that financial pressures are putting on profitability while effectively and efficiently monetising data growth and reducing vendor lock-in. This trend clearly shows that, in order to be sustainable in the near-future, operators networks will require the right amount of mobility, ultra high-speed networks, cloud computing, big data analytics and security.

Research into NFV performed by leading analysts firms confirms the development of NFV and reveals major market potential. In November, Mind Commerce estimated that the NFV market in 2014 will be worth $203 million, and will grow at 46 percent annually until 2019, when it reaches $1.3 billion. The research firm states that the chief domains targeted by early NFV deployments will be IMS services and the EPC. Last August ABI Research predicted a similar growth curve, with a potential $6 billion market for virtual networking by 2018. A new study from ReportsnReports.com forecasts that the NFV, SDN and wireless network infrastructure market will reach $5 billion by the turn of the decade, driven by rising global wireless capital expenditures and growing demand for high-speed mobile broadband. Wireless carriers will play a critical role in the SDN value chain, and that carriers will initially focus on southbound APIs and switch fabric, SDN and virtualization that will enable IMS optimization and realization of investment, and that by 2016 carriers will focus more on northbound APIs and create full development environments.

Network virtualisation allows operators to simulate network resources through SDN and NFV technologies that decouple, run and optimise different functions of the network.The industry is evolving from proprietary equipment networks to IT-based data centre networks that employ technologies such as software-defined networking (SDN), network function virtualisation (NFV), cloud-computing and big data analytics to provide a variety of converged services to consumers. NFV is highly complementary to SDN. Network functions can be virtualised and deployed without an SDN being required and vice-versa. According to ETSI, early NFV deployments are already getting underway and are expected to accelerate during 2014-15.Software-defined networking (SDN) and Network Functions Visualization (NFV) will drive changes in data security investment, according to a new report from Infonetics Research. Their Data Center Security Products report noted a shift in how organizations protect digital properties, including a 44 percent rise in the sale of purpose-built virtual security appliances. They anticipate a fairly significant revenue transition from hardware appliances to virtual appliances and purpose-built security solutions that interface directly with hypervisors, with SDN controllers via APIs, or orchestration platforms.

Rather surprisingly, communications service providers (CSPs) themselves, not vendors, are driving the development of network virtualization technologies. The potential to dramatically accelerate new service delivery, lower operating costs, and eliminate vendor lock-in has CSPs salivating and network equipment vendors scrambling. Vendors who sell proprietary network gear don't exactly welcome the thought of their intellectual property being replaced by standardized software running on commodity hardware. This has pushed the timeline for SDN and NFV further out, and prompted more than a few analysts to pull the hype card.The virtualization of service and control functions in the core network has been a first step in using cloud computing technology in the telco domain. However, for a full telco cloud implementation, virtualization needs to be complemented with a complete cloud platform and management system. This must include classical network management for legacy systems, plus virtualized network function, cloud orchestration and application management to achieve the full benefits of automated provisioning and elastic scaling of the network.

Driven by the promise of total cost of ownership reduction, wireless carriers are aggressively jumping on the NFV and SDN bandwagon, targeting integration across a multitude of areas including radio access network, mobile core, OSS/BSS, backhaul, and CPE/home environment.Telecom Italia has been among the tier 1 telcos driving the move to NFV. Along with AT&T, BT Group, Deutsche Telekom, Orange, Telefonica and Verizon, the company a couple years ago pushed network functions virtualization into the spotlight by creating an ETSI group to explore the technology. The key goals of the NFV Working Group are to reduce equipment costs and power consumption, improve time to market, enable the availability of multiple applications on a single network appliance with the multi-version and multi-tenancy capabilities, and encourage a more dynamic ecosystem through the development and use of software-only solutions.

Telefonica's UNICA platform is initially focused on virtualising signaling-related functions, including IMS (IP multimedia sub-system, DNS (domain name system), SMSC (short message service centre) and OCS (online charging system). The second phase will look at virtualising functions that carry traffic such as the core packet network. Telefonica's NFV programme is notably designed to "source different functions to different suppliers" and avoid vendor lock-ins. The company wants to design a virtualised network architecture that allows vendor interoperability.Among the many capabilities offered by UNICA is the idea of multi-tenancy (where the same basic solution works for multiple organisations) or NaaS (Network as a Service), using pre-installed templates to deploy virtualised equipment in real time and with integrated resource management.UNICA promises to offer real and permanent change for Telefónica's network transforming the company into a true Digital Telco.

Meanwhile ATT, has introduced its vision for the company's network of the future: the 'User- Defined Network Cloud.' AT&T claims their the cloud-based architecture is "a global first at this scale." The operator also announced the group of vendors that will work on implementing this strategy. The carrier expects its revamped architecture will accelerate time-to-market for technologically advanced products and services. Integrated through ATT's wide-area network (WAN) and using NFV and SDN, the architecture is expected to simplify and scale AT&T's network by separating hardware and software functionality, separating network control plane and forwarding planes, and improving functionality management in the software layer.This move to software-based telco environments will not only help incumbent providers become more agile and adapt to market trends and subscriber demands more effectively, but will open up the market to new players who may not have had such deep pockets needed to develop proprietary hardware. It will allow new carriers to quickly scale and compete, as they won't have to load up on costly central office equipment to get started.

For all you Telco Cloud builders something is floating up there : they call it Openstack !! OpenStack is a free and open-source software cloud computing platform that is primarily deployed as an infrastructure as a service (IaaS) solution. The technology consists of a series of interrelated projects that control pools of processing, storage, and networking resources throughout a data center, able to be managed or provisioned through a web-based dashboard, command-line tools, or a RESTful API. It is released under the terms of the Apache License. Founded by Rackspace Hosting and NASA, OpenStack has grown to be a global software community of developers collaborating on a standard and massively scalable open source cloud operating system.

So who is the big user of Openstack besides NASA? Well how about the Large Hadron Collider (LHC) project , which generates 1PB of data every second ?? CERN started using the OpenStack private cloud back in 2011 in the testing environment, upgrading more recently to the fifth version of OpenStack : the Essex release. Moving to a mammoth-scale infrastructure-as-a-service (IaaS) cloud based on OpenStack has helped the European Organisation for Nuclear Research (CERN) significantly expand its compute resources and support more than 10,000 scientists worldwide using the infrastructure to find answers to questions such as what the universe is made of. The big vision for CERN's private cloud infrastructure is to be able to scale up to hosting 15,000 hypervisors on the cloud by 2015, running between 100,000 and 300,000 virtual machines !!

A major strength of OpenStack is its ability to more easily enable hybrid cloud platforms - the combination of public and cloud operations that appeal to larger enterprises looking to combine the savings of commodity public clouds for some operations with the security and control of private clouds for other apps. Now that VMware has joined OpenStack consortium the plattform is expected to gain even more momentum as well among major telecom players operating their cloud subsidiaries. Today's Telcos are struggling to support ever-expanding demands from both consumers and enterprises, including the need to transmit and store increasing amounts of data. As the Internet of Things grows, these data throughputs will only rise alongside user requirements. According to pundits while other Cloud solutions can provide adequate support for boosted data traffic levels, OpenStack does so in a cost-efficient and elegant manner that other technologies just can't match. Competitive solutions are three to five times more expensive than OpenStack deployments . OpenStack's unique features and functionalities , such as allowing service providers to grow or shrink their offerings to match service peaks and no licensing fee , make it an ideal Telco Cloud platform.

DT is one company that has made the Open Source Cloud as the centrepiece of their global services strategy. Deutsche Telekom offers a portfolio of over 30 different cloud services that encompass infrastructure, developer environments, collaboration, business applications and security as a service. By employing the cloud service brokerage strategy, Deutsche Telekom has become the cloud partner that its millions of business customers are looking for. The loyalty and lifespan of those customers will be dramatically increased as they receive innovative, business-critical services from a provider they already know and trust.Deutsche Telekom enhances its set of cloud technologies with OpenStack. The open source cloud operating system makes it easy for software partners (ISVs) to integrate their cloud applications in the Deutsche Telekom infrastructure and its new Business Marketplace, removing technical obstacles. Deutsche Telekom's Business Marketplace is an online platform which will offer cloud services for small and medium businesses started in 2012. Their goal was to offer our customers a rich set of business applications out of our cloud. Most mobile cloud applications are sophisticated mash-ups of all sorts, including applications that incorporate core mobile phone and Telco capabilities, like location, presence, phone calendar, address book, and cameras.

The Business Marketplace consolidates innovative business customer solutions from DT partners. Usability is a particular focus: business customers can find, book and manage the cloud-based applications with a single click. In addition to detailed information on all products, users can also see other customers' ratings and try out the products free of charge before buying them. Business Marketplace also lets companies keep track of their application licenses and employees' access privileges at all times, as well as view their billing data. The users receive a single bill for all ordered applications on a monthly basis.

One of the most valuable services, that can enfranchise many thousands of new ASPs, is the 'Bill on Behalf-of' (BoBo) ability for them to be paid through their customers' phone accounts. Comprehensive payments and settlement and the associated business infrastructure, is a critical component of Mobile Cloud Computing and DT are leveraging this element . The operator also plans to contribute to the development of OpenStack and Deutsche Telekom has created a growing team of in house engineers that will work to harden and secure OpenStack. They have already started offering a security package for SMEs to protect them from viruses and attacks from the internet.

DT's aspiration is for €2bn in additional revenue by 2015 built on a track record of being among the first to bring leading-edge cloud innovation to customers in Europe covering domains like end user computing, enterprise networking and data center. DT firmly believes in developing an open ecosystem. It is not that the telco is afraid to innovate on its own behalf; indeed, it develops many of its applications and platforms in-house. It has, however, become very open to working with partners if they can either speed up time to market or add brand credibility.

Despite its many virtues OpenStack offers risks and rewards to telcos. On the upside, OpenStack allows telcos to more rapidly roll out cloud services at low cost.On the downside, OpenStack standards could allow end-customers to more easily migrate their cloud applications from one OpenStack telco to the next. Telco Cloud service strategies vary : some telcos are simply reselling Office 365 while adding some value-added services. Office 365 resellers include Bell Canada, KCN of the Netherlands, France-Telecom Orange, Telefonica, TeleiSonera, Telmex, Telstra and Vodaphone etc. Other telcos see OpenStack as a potential way to battle Amazon, Google and other cloud providers. Verizon's buyout of Terremark and CloudSwitch essentially positions Verizon against OpenStack-focused service providers. Its the wild west in the clouds !!

Even the network equipment vendors are clambering on the OpenStack bandwagon . Ericsson is now previewing a tweaked version of OpenStack that will run on its network iron. Ericsson is giving the same server consolidation pitch that has been common in the mainframe market for three decades, in Unix for a decade and a half, and for the past decade or so in the x86 server racket, to push OpenStack into its telco gear. At the heart of the Ericsson Cloud System is what the company calls the Cloud Execution Environment, which runs on x86 iron, of course. It uses the KVM server virtualization hypervisor championed by Red Hat to dice and slice virtual server instances on top of physical server blades and puts KVM and the workloads that run on its virtual machines under the control of OpenStack. And if Ericsson is on board then the Chinese vendors are not far behind !!

The fact that OpenStack was developed by a group of people who called themselves "open source" rather than a group that named themselves after a rain forest is probably irrelevant. The important thing is that, like Linux, it works robustly and is cheap and was built by people who knew what they were doing. The defining features of OpenStack is scalability and adaptability. As enterprises push for more open APIs into cloud platforms, both to ease the complexity of moving to the cloud and to prevent the dreaded vendor "lock-in," the pressure will mount on cloud operators of all kinds to embrace OpenStack.

As all Telco engineers know that in a typical mobile deployment, each base station serves all the mobile devices within its reach. Each base station has its digital component manage its radio resources, handoff, data encryption and decryption and an RF component which transforms the digital information into analog RF. The RF elements are connected to a passive antenna that transmits the signals to the air. Each base station should be placed in the geographical center of its coverage area. But even when such locations are selected, the mobile operators may have difficulty in renting the real estate, finding proper powering options, securing the location and protecting the equipment from weather conditions. Those cell sites carry with them a continuous stream of OPEX to address the high rental rates for real estate, electrical expenses, cost of backhaul for the cell site and security measures to protect the location from intruders.

Enter a novel architectural paradigm : C RAN !!! The basic premise of Cloud RAN is to change the traditional RAN architecture so that it can take advantage of technologies like cloud computing, Software-Defined Network (SDN) approaches, and advanced remote antenna/radio head techniques.C-RAN architecture is not bound to a single RAN air interface technology. In essence, conventional terrestrial cell site base stations are replaced with remote clusters of centralized virtual base stations which can support up to a hundred remote radio / antenna units. This is achieved by centralizing RAN functionality into a shared resource pool or "cloud" (the digital unit – DU, or baseband unit – BBU) which is then connected via fibre to advanced remote radio heads ("Radio Units" – RU) sited in different geographical locations in order to provide full coverage of an area. The radical concept can even use banks of x86 servers to connect cellular calls rather than traditional wireless base stations.

From a business perspective,C-RAN will deliver significant reductions in Opex and Capex due to reduced upgrading costs. A major reason for this is the aggregation and pooling of the DU computing power which can be assigned specifically where needed e.g. the load situation over time and space for indoor/outdoor cells, am/pm hours, weekday/weekend, and so on. As a result, single cells do not need to be dimensioned for peak hour demands, but rather the processing power can be pooled and assigned on an on-demand basis. The processing power savings achieved should also leave processing headroom for any further potential technology enhancements (e.g., LTE-A features) without the need for further CAPEX. C-RAN skips the need for a high-bandwidth, low latency (X2), synchronized interface between the geographically distributed base station because the computing resources of the multiple transmission points' BBUs are all located within the same hardware.C-RAN slashes capex because fewer BBUs are needed, which reduces opex because fewer BBUs means less energy consumption and diminished maintenance costs. The reduced energy consumption makes C-RAN a "green" alternative, with China Mobile estimating 71 percent power savings vs. traditional RANs.

Furthermore, interference management will also benefit from C-RAN network architecture as technologies like dynamic eICIC schemes will be enabled, especially in a HetNet deployment.Heterogeneous networks will require small cells to be independent, intelligent and ubiquitous to avoid the cross- interference mayhem, yet be in synch and orchestrated with macro cells (including Cloud - RAN topology).Small cells are poised to become the most commonly used node for cellular access in the next-generation HetNet. C RANs will likely take their place beside traditional base stations and emerging small-cell base stations as another tool for building cellular nets.The success of many new 4G network deployments will depend on the use of outdoor and indoor small cells to extend coverage and increase capacity in areas poorly served by macrocell networks. Operators are also considering proposals to deploy more efficient CloudRAN architectures requiring high speed CIPRI front haul links between remote radio heads and pools of baseband units.

According to Maravedis Analysts Cloud-RAN economics only be realized by harnessing standards to ensure interoperability and reduce cost. That, in turn, will create a whole new ecosystem, and operators must resist any attempts by their suppliers to hijack standards for software-defined networking or cell site equipment. Otherwise, this fledgling architecture will remain confined to a few pioneers with the resources to build their own ecosystems, like China Mobile. China Mobile, the world's largest carrier with 700 million subscribers, has been spearheading trials and plans to deploy systems as early as 2015. Japan's NTT Docomo said it will follow in 2016, and a third unnamed carrier is now preparing plans for C-RANs. China Mobile aims to lower the cost of C-RANs to less than $30 per LTE sector, down from about $10,000 two years ago. It will start a second round of trials later this year using servers equipped with PCI Express cards to handle baseband processing. Each card will pack four FPGAs using silicon cores, each FPGA capable of handling 12 LTE sectors.

As MNOs face rising CAPEX bills to meet mobile data demand combined with falling ARPU, they must explore radical new network designs. With Cloud-RAN, they can virtualize baseband processing functions for hundreds of sites on a server or base station hotel. By consolidating individual Base-station processing into a single or regional server farm Investments on Cloud Radio Access Network (RAN) Infrastructure are expected to exceed $6 Billion by 2020, according to a new report from SNS Research.

Distributed antenna technologies (DAS) will get a new lease on life, supporting coverage extension for C-RAN sites. This sector will open up $1.3bn in new revenues for antenna providers. Pure C-RAN faces many barriers, such as over-reliance on fiber to link sites and basebands and immature standards, but most operators will inch towards C-RAN using hybrid models. Development of microwave fronthaul technologies will be critical to improve the C-RAN business model . Whatever the challenges C-RAN offers a revolutionary approach to next-generation cellular networks deployment, management and performance. Fiber, needed for fronthaul, is crucial to C-RAN deployment, so it is no wonder that fronthaul is constantly brought up as Cloud RAN's biggest challenge. Fronthaul connects RRHs to the aggregated BBUs, with traffic then backhauled from the BBUs to the IP core or evolved packet core (EPC).

NTT DOCOMO, Japan's leading mobile operator and provider of integrated services centered on mobility, announced today it will begin developing high-capacity base stations built with advanced C-RAN architecture for DOCOMO's coming next-generation LTE-Advanced (LTE-A) mobile system. The new architecture will enable quick, efficient deployment of base stations, especially in high-traffic areas such as train stations and large commercial facilities, for significantly improved data capacity and throughput.Advanced C-RAN architecture, a brand new concept proposed by DOCOMO, will enable small "add-on" cells for localized coverage to cooperate with macro cells that provide wider area coverage. This will be achieved with carrier aggregation technology, one of the main LTE-Advanced technologies standardized by the Third Generation Partnership Project (3GPP). The small add-on cells will significantly increase throughput and system capacity while maintaining mobility performance provided by the macro cell.

For NTT DoCoMo high-capacity base stations utilizing advanced C-RAN architecture will serve as master base stations both for multiple macro cells covering broad areas and for add-on cells in smaller, high-traffic areas. The base stations will accommodate up to 48 macro and add-on cells at launch and even more later. Carrier aggregation will be supported for cells served by the same base station, enabling the flexible deployment of add-on cells. In addition, maximum downlink throughput will be extendible to 3Gbps, as specified by 3GPP standards.C-RAN is typically thought of as a large-scale urban macro solution, but the concept of pooled baseband serving n number of radio access nodes can apply to a variety of scenarios, such as small cell underlays (using micro RRUs), so-called Super Cells, and outdoor/indoor hotzone systems. These models, identified and defined partly through the NGNM Alliance, could prove an attractive way to introduce and develop C-RAN technology. Given the traditional RAN's coverage restrictions and limitations of transmission and reception signal support, the benefits of deploying a C-RAN infrastructure are clear.

The C-RAN, as a centralized, general purpose processing solution, enables the efficient use of network resources. Based on open-platform and base station virtualization, C-RAN provides an ideal architecture for LTE-A functionality as well as being complementary to next-generation SDN and NFV deployments. Many major mobile operators across the globe are preparing to incorporate the cloud into their existing RAN platforms. We anticipate that 2014 will move the C-RAN beyond the "cloud hype" as operators gain a competitive edge through integrating the C-RAN in their LTE-A migration.

Software Defined Networking or SDN is a technological approach to designing and managing networks that has the potential to increase operator agility, lower costs, and disrupt the vendor landscape.With SDN the network becomes a programmable fabric that can be manipulated in real time to meet the needs of the applications and systems that sit on top of it. SDN promises fully automated, application-aware and adaptive adjustments to bandwidth, compute power and storage with end-user visibility is what is needed to provide ultimate QoE to the mobile user connected an IP based Telecom network such as LTE .

The root cause of a network's limitation is that it is built using switches, routers and other devices that have become overly complex because they implement an ever-increasing number of distributed protocols and use closed and proprietary interfaces. By decoupling the network control and data planes, OpenFlow-based SDN architecture abstracts the underlying infrastructure from the applications that use it, allowing the network to become as programmable and manageable at scale as the computer infrastructure that it increasingly resembles. An SDN approach fosters network virtualization, enabling IT staff to manage their servers, applications, storage, and networks with a common approach and tool set.

In a SDN, the network administrator can shape traffic from a centralized control console without having to touch individual switches. The administrator can change any network switch's rules when necessary -- prioritizing, de-prioritizing or even blocking specific types of packets with a very granular level of control. This is especially helpful in a Cloud computing multi-tenant architecture because it allows the administrator to manage traffic loads in a flexible and more efficient manner. SDN allows network engineers to support a switching fabric across multi-vendor hardware and application- specific integrated circuits.

Most of the churn in mobile subscribers today is attributed to poor QoE (Quality of Experience). What's needed is an efficient and elastic system that adapts to the end-user traffic automatically and dynamically. The rapid adoption of 4G Mobile (LTE) necessitates uninterrupted availability of quality services 24/7 regardless of location or device. SDN has the capability to make this a reality. The Open Flow protocol allows the network to be programmed on a per-flow basis and thereby provides visibility at the user and application level. The capability to increase or decrease the bandwidth needed, for instance, by way of automated bandwidth signalling is one advantage. It can also adjust the number of VMs (Video Messages) and the associated storage needed proactively and dynamically with¬out any human intervention on an application basis.

Developing an SDN business involves the deployment of physical infrastructure, a network controller and a telecoms operating management system which combines operation and business support systems. The network controller is central to SDN with two main functions: virtual resource control and traffic management systems (TMS). The network controller can create a programmable, logical network that allocates resources within the physical network (access and core networks) in the most dynamic way without needing to know the actual infrastructure topology. In so doing, the operator can build the most appropriate virtual network offering multiple services.

SDN is not only an esoteric technology concept but a current reality: in 2012 Google announced that it had migrated its live data centres to a Software Defined Network using switches it designed and developed using off-the-shelf silicon and OpenFlow for the control path to a Google-designed Controller. Google claims many benefits including better utilisation of its compute power after implementing this system. Recently Japanese vendor NEC established a partnership with Portugal Telecom that will see the two firms collaborate on SDN (software defined networking) and virtualisation technology for data centers and carrier networks. The two firms claimed the agreement would enable both companies to test and assess the commercial feasibility and benefits of SDN implementation for carrier data centers , adding that SDN and network virtualisation have "exceptional potential".

According to Infonetics, telecoms plan to deploy SDNs and NFV by 2014 within data centers, between data centers, operations and management, content delivery networks (CDNs), and cloud services. In most cases, Telcos are starting small with their SDN and NFV deployments, focusing on parts of their network, in " contained domains " such as data centers, to ensure they can get the technology to work as intended.

Running in parallel Telco network architects believe that NFV (this term Network Function Virtualisation was coined by the European Telecommunications Standards Institute) will consolidate many network equipment types onto industry standard high volume servers, switches and storage, thus providing a new network production environment so as to lower cost, raises efficiency and increases agility. Network Functions Virtualisation can be implemented without the prescence of a SDN , although the two concepts and solutions can be combined to unlock greater value.

In the next decade SDN is big business.According to SDN Central (an independent market research community for SDN & NFV), the SDN market is expected to surpass $35 billion in the next 5 years. Adoption of SDN technology has accelerated in recent years from sales of $10 million in 2007 to $252 million in 2012.The emergence of the software-defined networking market is supported by growth in venture capital investment in SDN focused companies. Venture capital funding rose from $10 million in 2007 to $454 million in 2012.

NFV and SDN concepts are at the core of our strategy. These help us realise our future network vision, which is a mutli-service, multi-tenant platform where we can respond more quickly and efficiently to our customers' needs. With NFV, we're able to dynamically reroute traffic and add capacity without adding new boxes. With SDN, we're removing pre-defined physical limits of the network by shifting control from hardware to software. These allow the network to become simpler, more scalable. They also allow us to reduce costs significantly and more quickly address customer needs "

(John Donovan, Senior VP– Technology and Network Operations, AT&T)

In the past decade the telecommunication industry has been revolutionized by advances in three core technologies: photonics, microelectronics, and software. The emergence of high-speed optical transmission and switching plus 4G Ran is likely to fuel an already growing demand for interactive image communications, multimedia applications, and real time video services, including video conferencing, TV, and High-Definition TV. As such a deeper understanding of the architectures and protocols for broadband integrated services networks and the ability to highlight relevant performance issues becomes a critical skill.

Clearly the modern IP based Telecoms networks are more about software than hardware now. The BSS / OSS has become the brain of the network with its complex layers of middle ware to control and bill for traffic in high speed wireless data networks. In the current telecom market where the devices are smarter, the networks are accelerated and customers are well informed, the balance between OSS and BSS plays significant role in the quality of customer experience. OSS and BSS together enable the CSP's to consolidate, simplify and automate the operations.

According to IBM Tech Trends Report , mobile computing, cloud computing, social business and business analytics have gone beyond niche technology status and are now part of core IT focus. All of these technology trends require fast response times, vast stores of data, and a highly elastic backbone of networks and servers. The new software developed for clouds demands different kinds of code to take advantage of the flexibility of computing clusters.

Today's networks are facing the increasing pressures of mobility and BYOD, social media usage, and Big Data analytics. More bandwidth is required to support these trends, and IT is being challenged to reduce latency and deliver acceptable performance for cloud-based applications and services. One response to the escalating demand for faster, more efficient networks has been the emergence of software-defined networking (SDN). It's still early days for SDN, with adoption being confined largely to industry giants such as

Google. But as SDN matures, it could play a critical role in helping organizations define, provision, and manage their networks.

In the US there is a stampede for software talent. Companies and Universities battle to attract students with Maths skills to learn about software engineering or design exciting new platforms that leverage the Internet. Recruiters say the fiercest demand is for top-level, experienced workers with a few critical skills such as user interface, which involves designing the look and feel of a software application; mobile apps development, which entails programming for smartphones and tablets; and cloud computing software, which requires new kinds of code. Yet the Telecoms industry esp in the developing world is plagued by :

• An acute shortage of people with Science , Engineering and Technological competencies combined with essential management skills

• Unavailability of readily accessible information on trends and conditions in the labor market enabling correct career and learning choices and investment decisions

• Dwindling pool of technically competent and adequately prepared candidates from the youth market to take up available jobs in the science , information , technology sectors

• Lack of funding for focused training and education projects that will ensure continuous skills upgrading to keep abreast of the technological innovation

This " penury in competence " and ensuing structural unemployment has grave implications on the competitiveness of many Telco companies especially in developing countries . Many countries in Middle East and Africa have a high structural unemployment rate ie : the unavailability of skilled people for available jobs. Many unemployed are professional people with social sciences backgrounds that are worthless in the new InfoTech economy. Ofcourse it is a no brainer to recruit software engineers from the " software factory " nations like India if you can afford and wish to rely on expats. However this does nothing to address the problem of structural unemployment or rising joblessness among the youth in Middle East and Africa.

Software engineers apply the principles of engineering to the design, development, maintenance, testing, and evaluation of the software and systems that make computers or anything containing software work. Some of the basic competencies a typical Software Engineer learn are :

1. Study of the principles and practices of software engineering : software quality concepts, process models, software requirements analysis, design methodologies, software testing, and software maintenance. Hands-on experience building a software system using the waterfall life cycle model.

2. Problem-solving and program design using C++ : Introducing a variety of programming techniques, algorithms, and basic data structures—including an introduction to object-oriented programming

3. Software Testing and Quality Assurance : quality concepts, black and white box testing techniques, test coverage, test planning, levels of testing, the formation of a testing organization, testing-in-the large and special problems in object-oriented testing, documentation for testing, and inspections and walkthroughs as a vehicle for product quality

4. Oriented Information Systems : Investigation of different architectural strategies for building object oriented information systems. Develop familiarity with modeling, design, and implementation techniques used in the construction of object oriented information systems.

5. Software Metrics : Theoretical foundations for software metrics. Data collection. Experimental design and analysis. Software metric validation. Measuring the software development and maintenance process. Measuring software systems. Support for metrics. Statistical tools. Setting up a measurement program. Application of software measurements.

What is urgently required is an innovative industry-university-government collaboration to prepare math and science graduates for advanced study in software engineering, telecommunications, and satellite communication and provide them with convenient advanced degree programs This will ramp up the development of software and telecommunications engineering human resources, and help accelerate the development of ancillary telecommunications and software engineering industries in MEA Region. Telcos in MEA need to do more to accelerate the skilling software engineers by setting up Software Engineering Centres in the countries they operate instead of relying on universities.

Since 2008, mobile software and applications have moved from the sphere of cryptic engineering lingo to vital part of the essential marketing playbook for mobile industry vendors. In stock market terms, developer mindshare is one of the hottest "commodities" in the mobile business, one whose "stock price" has ballooned in the last few years. Platform vendors, handset OEMs, network operators, hardware vendors, and infrastructure providers all want to contribute to mobile apps innovation. All the Telco value chain players are now vying to win software developer mindshare, in order to add value on top of their devices and networks. Mobile application development and integration are at the forefront of the modern SOA story.

Key mobile trends in 2013 included the emergence of app stores, the HTML5-native debate, mobile back ends, RESTful services and open APIs. Widespread adoption of mobile applications is at the root of these changes in development, and it shows no signs of decline with the consumerization of IT. In general, mobile web development within an

HTML5 browser or web runtime is promising when it comes to market penetration, ease-of-use and cross platform support. Industry analyst Gartner Inc. predicts that more than 73 billion mobile applications will be downloaded in 2013, and that that number will nearly quadruple to 287.9 billion by 2016. Cloud mobile back-end services stand to become a key component of the application development ecosystem. By 2015, 80% of all mobile applications developed will be hybrid or mobile-Web-oriented .But how is the landscape of mobile developer mindshare looking today?

The recent report " Mobile Developer Economics & " contains new insights into the motivations of mobile developers . For instance , developers still consider fun and coding speed as very important even if developer mindshare is turning towards the appeal of monetization and reaching a large audience. The technical reasons for selecting a platform seem to be gradually becoming a less important selection criterion. For today's mobile developer, market penetration and revenue potential are the two most important reasons for selecting a platform. The most successful developers are those that extend apps to new markets, either to new geographies or different verticals. To some extent, these strategies rely on copying the recipe of an already established and successful business: these are apps that have been tried and proven in at least one market and are generally less risky options for developers.

Application developers are also increasing their demand for app store platforms, which provide a centralized place to buy, sell and manage their apps. More applications are designed with open APIs to enable application-to-application integration. As a result, the relationship between business and developer is shifting to give external developers more sway. However, the study shows several pain-points with mobile web technologies compared to native applications, namely issues with development environments, device API support and UI creation.

The goal of Telco API programs is to allow developers to take Telco services into new niches and use cases, and scale from hundreds to thousands of partners. Some of these new use cases will result in supplemental Telco revenue streams, some will facilitate customer acquisition, while others will subsidise ecosystem creation costs. APIs need the flexibility to allow developers to experiment with new use cases, and thus discover and satisfy unmet user needs. If Telcos allow and encourage developers to create locally-relevant differentiation on behalf of their subscribers, their fragmentation disadvantage could transform into the advantage of local presence.

Since there is no such thing as an "average developer", Telco API business models need to be designed to target one or more specific developer segments. To reduce friction and help developers discover new user needs and opportunities, Telco API business models need to subsidize experimentation and be designed for the ability to fail and retry cheaply. More specifically, if developers are charged based on Telco API usage, the app's business model must have a stable, usage-based income stream. By allowing free, small-scale usage of the API, Telcos permit developers to experiment with multiple business models, including free, until a sustainable, workable business model can be found.

Telco APIs will always be at a disadvantage versus players with global reach, if positioned in direct competition to native platforms or Internet companies. To be successful in API initiatives, Telcos need to consider developers as value-added resellers, and therefore design their API propositions for win-win outcomes. In other words, the business models of Telco APIs need to be aligned with the business models of developers. It is important to note that the same ecosystem economics that work for Telco APIs and app developers can be applied to other types of partners and service providers, such as Mobile Virtual Network Operators (MVNO) or machine to-machine (M2M) initiatives. MVNOs can build ecosystems around the distribution business layer. App developers can build ecosystems around the service layer. And M2M companies, meanwhile, can build ecosystems around the connectivity business.

But be warned : The Mobile App economy displays one of the key tenets of The Disruptive Technologies Model : which postulates that.... the pace of technological progress generated by established players inevitably outstrips customers' ability to absorb it, creating opportunity for up-starts to displace. This new theory provides a useful gauge for measuring not only where competition will arise but also where, in an industry's shifting value chain, the money will be made in the future. In the 4G world the dominant firm-level MNO value chain is ripe for unbundling in response to accelerated product/services evolution. Future success within the industry will go only to those Telco players with strategic foresight to "skate to where the money will be" : solo or via partnerships !!!

---♠---

Telco Challenge# 8 : The Quest for Connected Machines

Telco Global Connect

The convergence of efficient wireless protocols, improved sensors, cheaper processors, and a bevy of startups and established companies developing the necessary management and application software has finally made the concept of the Internet of Things (IoT) mainstream. The number of Internet-connected devices surpassed the number of human beings on the planet in 2011, and by 2020, Internet-connected devices are expected to number between 26 billion and 50 billion.

As with many new concepts, IoT's roots can be traced back to the Massachusetts Institute of Technology (MIT), from work at the Auto-ID Center. Founded in 1999, this group was working in the field of networked radio frequency identification (RFID) and emerging sensing technologies. The labs consisted of seven research universities located across four continents. These institutions were chosen by the Auto-ID Center to design the architecture for IoT.

The vast majority (80%+) of IoT connections will occur on unlicensed wireless frequencies due to cost and battery life advantages. Whereas cellular IoT connections are expected to grow at a nearly 20% CAGR for the next several years, various personal area network (PAN) wireless connections (Wi-Fi, Bluetooth, Zigbee) into M2M (machine-to-machine) end markets should grow closer to 30%. Aside from significant growth in PAN-based chipsets, some experts view the microcontroller and sensor opportunities as meaningfully positive with industry forecasts of single-digit growth potentially being too pessimistic.

In any case the IoT $ opportunity is not lost to Telcos and the senior execs are making noises now while others are forging ahead. Some Operators have already taken the lead in supporting such global service launches in early market categories such as automotive, health and consumer electronics. With the emergence of new products in adjacent categories such as healthcare, wearables and consumer electronics the importance of the ability to support large-scale global deployments is likely to accelerate.

" From an ATT perspective, we are providing an end-to-end (E2E) platform that enables a unified experience across a wide range of devices and one-off capabilities in the market today. There's a lot of noise in the marketplace that can make it complicated for a consumer or a business to make sense of this notion of the Internet of Things. It is important for us to show real benefits, whether you're a consumer or business " (Kevin Petersen, ATT Digital Life President)

" In the emerging Internet of Things (IoT) space telecoms operators will likely continue to provide connections into consumers' homes and power the different screens. But to remain relevant within the connected home landscape , operators will need to find new ways to interact with the things people care most about in their homes – things that help them stay comfortable, help keep them safe and help them save energy. (Chris Borros Nest Labs)

"It's (IoT and Digital Life Platform) not just a differentiator, it's an imperative for success. Nearly every CIO I talk to has security as his or her number one concern. What IoT can do for businesses is so exciting, but customers want to know their data is secure," (Ralph de la Vega, president and CEO of ATT Mobile and Business solutions)

Acording to Cisco IBSG there are several barriers, however, have the potential to slow the development of IoT. The three largest are the deployment of IPv6, power for sensors, and agreement on standards. It is important to note that while barriers and challenges exist, they are not insurmountable. Given the benefits of IoT, these issues will get worked out. It is only a matter of time.

The challenge for operators is to find a business model that delivers value for customers and is profitable. A significant part of their challenge is determining what an IoT network architecture and business operation should look like. These dilemmas need to be resolved quickly because, within the next five years, serving the IoT market will become the critical mission for any communications service providers.As industries such as automotive, utilities, transport and logistics feel the competitive pressure of IoT, the scramble for partners to help them will accelerate. One of the principle capabilities these companies will seek of their partners will be their ability to deliver complex solutions quickly.

Telcos are busy assessing alternative options for new Low-Power Wide-Area (LPWA) networks, since the connectivity revenues alone from these networks are exceedingly low: typically USD2–3 per device per year, but below USD 1. With such low revenues on offer, Telcos are wondering whether they should invest in LPWA networks. GSMA believes that While connectivity will underpin the development of the Internet of Things, to avoid becoming commoditised, mobile operators must leverage their networks' potential to provide value added services and build what could become a US$422.6 billion industry. In the case of the overall market revenue of US$422.6 billion, the majority of these revenues are to be derived from the 'Service Wrap'.

The 'Service Wrap' comprises the service that the end customer pays for that relies on the underlying connectivity, and operators are investing in building new capabilities that improve their offering to IoT service propositions. Examples include horizontal capabilities such as remote provisioning of IoT devices, building platforms that allow for management of business rules, reporting, support for Application Programming Interfaces (APIs) and the management and presentation of data. Moreover, 'Big Data' analytics is set to become a key part of IoT services in the future, with operators increasingly looking at ways to analyse data from various sources and create new service lines.

Telefónica has launched a modular internet of things platform called Thinking Things, which consists of stackable modules for a variety of purposes.There will be many sensors, actuator modules and so on to come, but the first manifestation of the new platform is an "ambient kit pack" that includes a communications module with an embedded SIM, a module for measuring air temperature, humidity and ambient light, and a battery module that can be charged via microUSB (the battery modules,which can charge 1,000 "communications" per charge,) can themselves be stacked.

This will let users remotely control the temperature, lighting and humidity of their home or office, though that only applies to lights, heaters and humidifiers that are plugged in at the wall, rather than fixed units. That use case will probably also require the smart plug module that Telefónica will release early next year, allowing users to turn devices on and off, dim lights and measure energy usage. Telefónica has released modules for sensing presence, impact and audio, and notifying the user via LEDs.

"This is a major step in Telefónica's journey into the internet of things," the company's director of industrial internet of things, Francisco Jariego has said. "Our aim is for Thinking Things Open to become an open ecosystem in which any object or device can be connected to the internet."

The IoT will increase the range of services, each requiring varying levels of bandwidth, mobility and latency. For example, services that are related to public safety or personal safety will generally require low latency, but not high bandwidth per se. alternatively, services that provide surveillance might also require high bandwidth. Due to the differing level of service demand, mobile networks may need the ability to identify the service which is generating traffic and meet its specific needs. For example, alert services related to public safety or personal health would require a higher priority compared to metering information, which is a normal monitoring activity.

For every Internet-connected PC or handset there will be 5-10 other types of devices sold with native Internet connectivity. These will include all manner of consumer electronics, machine tools, industrial equipment, cars, appliances, and a number of devices likely not yet invented. In the world of IoT, even cows will be connected. A special report in The Economist titled "Augmented Business" described how cows will be monitored .Sparked, a Dutch start-up company, implants sensors in the ears of cattle. This allows farmers to monitor cows' health and track their movements, ensuring a healthier, more plentiful supply of meat for people to consume. By the way on average, each cow generates about 200 megabytes of information a year .Bottom Line : Jump on the IoT bandwagon for the right reasons. Be prepared.. its not an trial initiative but a way of life !!

Next generation utility meters – or 'Smart Meters' – are a good example of the transformative potential of M2M technology. They will empower consumers by providing them with feedback on their energy usage, helping them to monitor, manage and - should they wish - reduce their energy consumption. Smart meters will also help reduce or end estimated readings, and make it easier for consumers to change tariffs and switch between suppliers, increasing market competition.

For the neophytes a smart meter is usually an electronic device that records consumption of electric energy in intervals of an hour or less and communicates that information at least daily back to the utility for monitoring and billing purposes.Smart meters enable two-way communication between the meter and the central system. Unlike home energy monitors, smart meters can gather data for remote reporting. Such an advanced metering infrastructure (AMI) differs from traditional automatic meter reading (AMR) in that it enables two-way communications with the meter. Smart meters are invariably part of an integrated program that pays for itself through reduced theft of electricity, energy savings,and operational efficiencies.

Equipping households and business with smart meters and collecting data on energy use enables better balancing of the network and a live view of grid activity, including technical failure. Granular, aggregated and analytical data generates information for better network planning and optimization to reduce losses.Data from the meters can be analyzed to detect unexpected consumption patterns, which can also indicate potential theft.Tamper controls can be built in to smart meters to alert utility companies to power theft.

The world's largest smart meter deployment was undertaken by Enel SpA, the dominant utility in Italy with more than 30 million customers. Between 2000 and 2005 Enel deployed smart meters to its entire customer base. These meters are fully electronic and smart, with integrated bi-directional communications, advanced power measurement and management capabilities, an integrated, software-controllable disconnect switch, and an all solid-state design. They communicate over low voltage power line using standards-based power line technology. Consumers are able to access their energy consumption and billing data online, and new methods of payment will be possible.

The overall revenue opportunity in smart electric metering is substantial, amounting to nearly $57 billion over the coming years, Navigant Research forecasts annual revenue will grow only fractionally, from $5.2 billion in 2012 to nearly $5.3 billion in 2022, with a compound annual growth rate (CAGR) of 0.1%. The Western European and Asia Pacific markets represent healthy growth potential, both in installation of new smart electric meters and upgrades to existing smart meter technologies. Driven primarily by China, the forecast penetration rate in Asia Pacific will reach nearly 70% by 2022. Growth in Europe is due largely to the EU smart metering policy which calls for 80% of households to have smart gas and electric meters installed.

The World Bank estimates that Africa needs to build an additional 7,000 megawatts of new generation capacity each year to meet suppressed demand, keep pace with projected economic growth and provide additional capacity to support the rollout of electrification.A push to introduce smart meters could help to relieve Africa's stressed power supply and bridge this investment gap in the region's electrical infrastructure. Put simply, installing a prepayment smart meter guarantees that cash can be collected from a customer. This could provide the necessary missing link to secure revenues, thereby attracting global investors in new generation capacity and allowing investment in other critical energy infrastructure.

Modern energy systems built on top of smart meters offer enormous opportunities to nurture smart grid (a smart grid is a modernized electrical grid that uses communications technology to gather consumer data in an automated fashion to improve the efficiency, reliability, economics, and sustainability of the production and distribution of electricity), smart home and e-health solutions, all currently in their infancy. As a result of these new solutions, there's a need to scale effortlessly to cope with new and unpredictable innovations in services and products, and thus data traffic. Cellular is a technology that has proved that can scale economically future-proof smart energy systems. It's secure, open and interoperable, and cellular's clear roadmap and continued development will support the Smart Metering deployments as the technology continues to evolve to focus on high-speed capabilities and rich data services.

" From an ATT perspective, we are providing an end-to-end (E2E) platform that enables a unified experience across a wide range of devices and one-off capabilities in the market today. There's a lot of noise in the marketplace that can make it complicated for a consumer or a business to make sense of this notion of the Internet of Things. It is important for us to show real benefits, whether you're a consumer or business " (Kevin Petersen, ATT Digital Life President)

A D Little point out rightly that in a saturated market, smart Telcos are now starting to move from mass-market products to more sophisticated and vertically integrated solutions as a means of addressing declining revenues in the residential telecoms market. However, their existing core competencies seem almost predestined for expansion into certain areas of Smart Grid allowing them to gain a foothold in the electricity market. The existing automated meter reading business has not tempted telcos, since the low level of data traffic on SIM-enabled meters makes average revenue per user (ARPU) unattractive. To address this, the fundamental business model for automated meter reading needs to be broadened to include billing and management services for the energy market and the end-customer. Moving from a connectivity-oriented business, which represents only 10–15% of the value generated in this field, to a service provision model would allow operators to extend their share in the value chain up to 60–70%. As telecoms operators are currently developing their skill sets and platforms, it seems it will be only a matter of time before they move en masse into this highly attractive area.

Deutsche Telekom is counting on getting an extra €1 billion in revenue annually by 2015 from networking services targeted specifically to the energy, automotive, health and media industries. Deutsche Telekom's plan is to sell a fixed-price service that includes installation and operation of a communications box that transmits the smart-meter data to a secure data center every 15 minutes, computes it and then forwards it to the utilities. It also provides utility customers with the ability to view their power usage via a secure Internet site.

The UK's Smart Meter Implementation Programme is a major national infrastructure project that will involve the roll out of 53 million gas and electricity meters across the UK by 2020. Telefonica's proposed communications solution for the smart metre network is based on its existing cellular rollout, supported by am IPv6 based wireless mesh solution which will connect meters in areas without cellular coverage. The system is based on the 6LoWPAN initiative founded by an IETF working group dedicated to pushing IPv6 over Low power Wireless Personal Area Networks under the idea that "the Internet Protocol could and should be applied even to the smallest devices." The IPv6 mesh network can then be established over any available carrier, including but not limited to, 2G/3G cellular, wifi, power line IP, and Bluetooth.

Telcos face an interesting business opportunity and they can become one of the few players able to gain access to a share of the entire energy management market. Their ability to utilize their existing network infrastructure and enabling capabilities to manage critical applications much more effectively will be a strong point of differentiation compared to application providers without their own network infrastructure. In CRM and billing, there are many similarities between telcos and utilities, particularly when smart meters and flexible tariffs are taken into account. Moreover, existing customer relations can be exploited as a sales channel, and bundling electricity metering with other telco products would be feasible (e.g. home or office automation, remote-management-over-mobile devices etc.). However, telcos would need to develop competencies in grid connection, utilities tariffs etc. The greatest challenge for telcos is to formulate a robust strategy and appropriate processes that allow them to act quickly and flexibly in response to the changing economic, technological and regulatory environment.

The bottom line : smart meters can create a "virtuous circle" linking energy companies, investors, customers, communities and regulators . With prepaid smart meters in place in homes and businesses, investors will be reassured that cash can be collected from customers to pay for investment in generation. Governments and regulators can fulfill their targets of controlling and optimizing energy consumption. Energy companies can improve customer service, run efficient network operations and dynamically manage supply and demand. And Telcos can utilise their vacated GPRS / Edge channels as their smartphone customers clog up the 3 G LTE bearers.

Browsing around the GSMA Connected Living pavilion is always inspiring because it is what mobile will do for our future.The Connected Car section exhibits exciting developments in the auto / mobile industries partnerships. After all we love our cars and spend so much of our life commuting.For the neophytes a Connected Car is a car equipped with on board localization and communication technologies,internet access, and usually also with a wireless local area network.This allows the car to share internet access to other devices both inside as outside the vehicle and to interact with other vehicles and infrastructures.

" We believe cyber security and approval from regulatory authorities will be key challenges when it comes to connected cars, but says both Nissan and Renault know exactly where they are headed with "autonomous" cars (not driverless). (Carlos Ghosn CEO at Renault – Nissan Alliance "

Examples of Connected Car technologies are: In-vehicle navigation system with GPS and TMC for providing up-to-date traffic information, Adaptive cruise control (ACC), Lane departure warning system,Lane change assistance system, Collision avoidance system (Precrash system),Intelligent speed adaptation or intelligent speed advice (ISA) system, Night Vision system, Adaptive light control system, Automatic parking system,Traffic sign recognition system,Blind spot detection system, Driver drowsiness detection system,Vehicular communication systems,Electric vehicle warning sounds used in hybrids and plug-in electric vehicles, etc.

The opportunity for the Connected Car market is huge, both in terms of revenue and benefits, such as customer loyalty.The global connected car market will be worth €39 billion in 2018 up from €13 billion in 2012, according to new market forecasts.The market is close to the tipping point where connectivity in cars will become a mass market.As 4G LTE networks reach people , virtually every auto manufacturer is working toward a connected car that takes advantage of next-gen data speeds, from voice-controlled apps and infotainment to advanced diagnostics.To build this new market, the mobile and the automotive industry will need to work in collaboration to surmount the challenges and deliver its promise.The industry is gearing up for a significant shift that will leave the landscape changed forever : whether you are a mobile network operator, automaker, software developer or hardware vendor there are huge opportunities on offer.

In 2011,Machina Research conducted a market sizing study (Machine-to-Machine (M2M) Communication in the Automotive Sector (2010-20), which forecast major growth in both telematics and infotainment services between 2010 and 2020. Automakers indicated that:Telematics and infotainment will be offered across vehicle brands, with a critical mass on embedded solutions :Tethered solutions will continue, with a focus on providing upgradable solutions for technology and, hence, the higher bandwidth services, i.e. infotainment, high bandwidth apps (music & video) : Embedded solutions will continue for vehicle-centric,high-reliability and high availability apps (such as eCall and bCall) :Infotainment and video services are expected to grow exponentially.Machina's forecast for global wireless traffic generated by embedded mobility in the automotive sector shows entertainment and internet access driving an exponential increase in data traffic.

Mobile operators are seeking to better understand the auto industry's requirements with respect to: How in-vehicle services, and their connectivity requirements,are evolving.How to enable all appropriate connectivity options for services.Greater understanding of these two aspects will facilitate the development of tailored approaches and services to support telematics and infotainment,in line with the underlying needs of automakers.

Moreover,cross industry collaboration will be required to overcome some existing ecosystem barriers. Mobile operators are particularly interested in fostering this joint collaboration in areas such as: Operational improvements on how to optimise data usage, common requirements for services and improving service delivery for different types of connectivity:new means to foster telematics and infotainment business development, such as through joint application programming interfaces (APIs), apps development and location-based services.

GSMA,along with a group of leading mobile operators,has already finalised the market requirements for the development of standardised embedded SIMs and for the remote management of SIMs.This has paved the way for the implementation of a worldwide-embedded SIM standard, reducing fragmentation and driving scale for 'connected' devices across various industries, including automotive, consumer electronics, healthcare and utilities.The goal of this initiative is to enable remote SIM management,helping drive global momentum for new, innovative and cost effective connected devices that will enhance daily life, while retaining the security and flexibility of current SIM card form factors.

ATT was the first to enter the market with a proprietary, single global SIM platform giving automotive, consumer and M2M equipment makers the ability to work through a single carrier to wirelessly enable and connect products.Announced last year, AT&T's single SIM platform delivers built-in access to wireless and data networks throughout most of the world, with service in more than 200 countries and access to more than 600 carriers worldwide.ATT spearheads two major initiatives to lead innovation in the connected car market – a first-of-its-kind connected car center in Atlanta, called the ATT Drive Studio, and a modular, global automotive platform called AT&T Drive.The ATT Drive Studio integrates ATT solutions across multiple companies and serves as a hub where ATT can respond to needs of automotive manufacturers and the auto ecosystem at large.ATT Drive Studio showcases end-to-end solutions that ATT and its contributors can provide automotive manufacturers around the world.Significant ecosystem players are committed to the Drive Studio and will work alongside ATT, including Accenture, Amdocs, Clear Channel's iHeart Radio, Ericsson, Jasper Wireless, Synchronoss and VoiceBox.

A number of options exist to connect a vehicle, including: Embedded: Both the connectivity (modem and UICC) and intelligence is built directly into the vehicle Tethered: Connectivity is provided through external modems (via wired, Bluetooth or WiFi connections and/or UICCs), while the intelligence remains embedded in the vehicle. Integrated: Connectivity is based upon integration between the vehicle and the owner's handset, in which all communication modules, UICC, and intelligence remains strictly on the phone. The human machine interface (HMI) generally remains in the vehicle (but not always). Each of these different connectivity options relies upon different mechanisms for linking the car to cellular technology.As the M2M market grows, so does the maturity and intelligence of M2M cloud platforms, enabling intelligent devices, back-end systems and cloud platforms to seamlessly integrate. A M2M cloud platform could bring about a global solution for managing connected devices across different networks and interfaces. This is attractive to the end-user facing brand (typically the automaker) as it enables the performance of the device on the network to be visible and troubleshooting processes to be performed.

A key factor driving the Connected Car is that connectivity will be necessitated by regulatory mandates such as the European Commission initiative eCall, which calls for a system to be fitted to all new vehicles by 2015, meaning emergency services will automatically be contacted and given the vehicle location in the event of a serious accident. Automotive OEM manufacturers across the board are fully prepared for the eCall legislation, although to what extent and when it will finally be implemented is unclear.There is no doubt throughout the industry that the connected vehicle will provide significant advantages in terms of life saving solutions and stolen vehicle tracking.

So what do we envisage for the Connected Car over the next ten years? Futurist Ian Pearson (@timeguide) see the vehicle developing into a fully personalised, virtual environment with intelligent automation, creating a totally new relationship between the vehicle, the driver, and the passenger. As you get in, the seat will automatically move to your preferred position, as instructed by your phone. Even fabrics and other interior surfaces will be able to adapt their appearance and textures electronically to your taste. Heads-up displays let drivers keep their eyes on the road.

Many people will wear video visors that overlay data onto the field of view, making augmented reality a part of everyday life, and changing the appearance of everything around us, including car interiors and the world outside.In-car sensors will recognise and highlight points of interest and dangers ahead. Passengers will see an electronically enhanced world too, with information overlaid into their view, in addition to games, video entertainment and web access.Some of this data will come from the car and some from apps on their phone.

Further in the future, cars will come to you. They will take you where you want, and then you can just abandon them to go off to serve someone else. They will in effect offer a comfortable and socially inclusive form of public transport. This could even lead to buses disappearing from our streets WOW..exciting times ahead as the mobile and automotive industries work hard to connect cars and networks for our pleasure and their profit !!

M2M is rapidly approaching a tipping point, a perfect storm of converging trends that creates the potential for fast and enormous growth. The democratization of device and service development is picking up momentum. You don't have to be an experienced engineer working for a top-notch technology company anymore to start building M2M solutions. The modularity that hardware and software developer tools introduce is lowering the barriers for developers and even end users to innovate. the emergence of open, ubiquitous general purpose technologies will make it possible to develop, launch and maintain new applications with dramatically lower needs for capital and lead times. This would potentially be an open-source technical stack, analogous to the LAMP (Linux/Apache/MySQL and one of Perl, PHP, or Python) stack ubiquitous on the Web.

Analysts believe that Machine-to-machine (M2M) is often portrayed as a nascent industry sector. However, operators are already generating strong revenue in this sector, amounting to USD10 billion worldwide in 2013, and increasing to USD88 billion by 2023. Future growth opportunities will be realised in emerging regions as applications are tailored to local markets and the cost of solutions declines. All the market signs show that the industry is ripe for an M2M ecosystem to emerge in the coming years. Whoever takes the lead, whether it is telcos or other players, will have a strong advantage in building network effects and locking in users and developers. According to VisionMobile's M2M Recipe these are the key ingredients to the M2M banquet :

M2M is rapidly approaching a tipping point : Modular hardware and software components are lowering the barriers for developers and even end users to innovate, paving the way for new entrants in the market. As communications specialists, telcos are in pole position to take advantage of the unprecedented growth that innovation promises, but only if they play their cards right.

Ecosystem economics have proven to be a source of decisive competitive advantage in recent digital revolutions such as the PC, internet and smartphones. Most recently, Apple and Google used ecosystem principles to open up a whole new mobile computing market and created tremendous value in vibrant, large app ecosystems. Incumbents who missed the opportunity to leverage ecosystem economics have lost their market positions to the ecosystem driven newcomers.

It is only a matter of time before someone applies the same ecosystem principles to disrupt the M2M market. M2M is today where the app market was in 2008. Many of the roadblocks in pre-2008 app development can be found in M2M today: high market fragmentation, the lack of direct access to customers, complex, tightly controlled and expensive (for the service developer) distribution channels.The source of possible explosive growth are new users, for whom current M2M solutions are too complex, expensive, or both. Making better, more advanced solutions for existing customers and system integrators will not attract those non-consumers. It is complexity and rigidity, not lack of performance that keeps new users from getting on board.

Non-consumption can best be addressed by leveraging a large and diverse set of developers. We don't know which M2M applications will emerge; indeed, it is fundamentally unpredictable. The app economy clearly showed that when you empower third party developers to experiment, they will find ways to create value, often in unexpected places.The advantages of empowering external developers are clear. The likelihood of uncovering "killer apps" is greatly increased, the risk of failure is off-loaded to a large amount of companies and individuals and telcos can unlock a level of investments far greater than any single company could afford.

The owners of successful platforms connect users and developers. It is the network effects resulting from their interaction that will ultimately create unprecedented growth.A successful platform leverages three key control points: service creation, distribution and consumption. Each of these control points needs to be designed to reduce friction and amplify network effects.A platform strategy is only defensible if the platform owner can capture some the ecosystem's value. The trade-off between stimulating growth and capturing value can be managed by subsidizing or commoditizing ecosystem participants and by designing the platform to drive the telco core business.

" In the emerging Internet of Things (IoT) space telecoms operators will likely continue to provide connections into consumers' homes and power the different screens. But to remain relevant within the connected home landscape , operators will need to find new ways to interact with the things people care most about in their homes – things that help them stay comfortable, help keep them safe and help them save energy. (Chris Borros Nest Labs)

Telcos have a strong incentive to own the M2M platform, rather than leave it to external companies. Only the platform owner can effectively use the tools of subsidies, competition stimulating openness and value-capturing closedness. Telcos can add value by making it easier to use their connectivity and providing more "M2M-friendly" interfaces – often described as managed connectivity. Beyond this, they can look to create and participate in the service enablers market for developers and application providers to easily identify, authenticate, provision, and maintain their device fleet;to update and rollback software on the devices and enable them to deploy processing logic into the "Internet of things" in order to render the system more robust, distributed, and autonomous.

AT&T's Digital Life initiative has packaged a set of home automation solutions around security, access, energy and water. This makes it easy for consumers to find and select solutions that suit their needs. On the one hand, because ATT only sells products from carefully selected vendors, consumers can be assured of a certain quality standard. On the other hand, as ATT manually selects solutions and presents them as its own, the company doesn't give the user much choice (tens of products at most) and makes it difficult for product developers to join the program. The M2M Marketplace of Deutsche Telekom is an early example of what an app store equivalent for M2M might look like. Vodafone is able to provide a global SIM to support your M2M capabilities and applications across the globe. By providing a global SIM that is pre-provisioned and ready to use, they can significantly reduce the complexity of installation, distribution and deployment of M2M solutions.

Telcos that have assembled the appropriate teams and resources will be poised for greater success as the M2M market begins to grow. As operators in developed markets have learned, it takes 18 months or more to organise the various aspects of an M2M business.Telcos now have the opportunity to reap the first mover advantages they didn't seize as the smartphone disruption unfolded, to take up a position of control and to avoid the need to respond to a commoditization scenario. The time to act is now.YOU SNOOZE YOU LOSE !!

M2M market (connecting devices other than phones, laptops and similar consumer devices is the new gold mine for network operators seeking data revenues. According to the international research firm Gartner, M2M is one of the Top10 mobile technologies to watch. Berg Insight predicts the number of cellular connections used for machine-to-machine communication to grow at a compound annual growth rate (CAGR) of 25.6 percent to reach 187.1 million worldwide by 2014. In the same timeframe, M2M's share of the global cellular network will rise from today's 1.4% to reach 3.1%

Telcos must understand that while mobilizing the machine is based on the same cellular network as that used for voice services, in most other respects the offering differs. Characteristics such as time of usage, frequency of usage, file size, and customer base will all be different. Consequently, rate structures, sales channels, service levels, and technical support must all be re-designed with embedded mobile in mind. M2M (sometimes called Embedded mobile) is likely to generate a lot of additional data traffic, so it will be an important potential source of revenue and growth. Particularly if some of the data traffic can be steered toward the network's off-peak periods. Network infrastructure is a sunk cost, and operators have predictable periods when they know the network will be under-utilized. According to Machina the market that is addressable by CSPs is $339 billion but of that only $5 billion will be about connectivity. But connecting it all, and connecting it at a level of service that CSPs need to guarantee must happen as stage one. As with people, CSPs need to choose their opportunities. Low ARPU connectivity may be a model for one CSP, value added services might be the model for another.

Amongst the pressing issues are standards adoption, cloud based service delivery platforms and security. With M2M network operators will probably not own the customer relationship as they do with voice. M2M applications are too complex and specialized for most companies to develop and deliver on their own. Instead, different players will need to collaborate to provide a total solution. For example, creating a building-automation system involved a range of different actors, including the chip and RF-chain vendors, the device assemblers, the operating system vendors, the network operators, and the application developers.Sample M2M applications include but not limited to :

1. Smart Metering : Smart Metering has become an essential part in the public utility industry. With government regulations and cost increase of nature resources (crude oil, coat, nature gas,etc), public utility companies have to implement an automatic real-time meter reading process to save cost and time. The real-time utilities usage information not only enables providers to allocate resource more efficiently but also allows customers to obtain more accurate billings.

2. Sales and Payment :Electronic payment system becomes the essential portion of people daily livings. More and More point-of-sale (POS) terminals, nowadays, are embedded cellular wireless modules in order to operate independently anytime and anywhere. Cellular wireless POS terminals can also feedback real-time sale information to vendors for resource allocation and sale & marketing analysis.

3. Security : The concern of safety is raising sharply. The demand of alarm systems to be installed in private residential, commercial and public locations grows exponentially. Cellular technology is applied to organize the usage of alarm systems according to different desires. Law-enforcement units, emergency organizations and private individuals can react more effectively upon the real-time alerts of various occurrences.

4. Healthcare: With cellular technology enable device, patients no longer have to be present in physician clinics to obtain medical advise. Physicians now are able to monitor their patients' condition continuously regardless of the local of the patients. Unnecessary clinical visits can be eliminated. Healthcare system will become more effective and efficient.

5. Telemetry : Shipments of medicine or food require the observance of certain temperature ranges or a complete documentation of the cold chain. Sensors inside the containers measure parameters such as temperature and humidity and transfer them to a central database using M2M technology.

Telcos are sitting on a gold mine of core capabilities and assets, critical for M2M success. In addition to just network connectivity, they have outstanding expertise in large scale service delivery with high reliability and global reach, plus an arsenal of partners for content, apps, specialized solutions and services, integration, as well as terminal and module vendors.While the price of devices has been steadily decreasing, mobile operators all over the world have been driving growth in the M2M market over the past several years as we can see by rapidly increasing numbers of M2M subscribers and higher revenues.

Telia Telenor is the Telco industry's poster boy in providing M2M applications based on their of horizontal, multi-tenant platform for M2M .The design of their Telenor Objects' software platform is a layering concept which provides a middleware layer to help different kinds of devices, networks, and applications to interoperate. This facilitates open interfaces for data capture, data pre-processing, device management, and information exchange with other systems. Telenor Objects is based on five key elements : the brand and channels to market, the technology, the partners, a managed-service business model drawing on the telco heritage, and open-source software.The complete system provides a range of reuseable capabilities, including a GUI for users, developer APIs, and a device library. The system core is a software platform which implements a secure message-exchange and device management system.

Sprint launched its Emerging Solutions Group in 2009. Some of the M2M offers include remote monitoring to keep track of Alzheimer's patients and providing news and weather updates for digital signage systems. They are also involved in remote monitoring and control of equipment, primarily SCADA units for oil wells and waste water, as well as security, utility meters and appliances. Some of their partners include Grid Net, Landis+Gyr, and Ford Motor Company.

Verizon is targeting enterprise clients with wireless ATMs, electronic medical records, RFID-tagged supply chain applications, video surveillance, stolen vehicle tracking, monitoring of water, gas and electric distribution, control electronics, DVD kiosks and vending machines.

Orange is focusing on vehicle tracking and online monitoring and reporting, asset management, wireless CCTV safety cameras, satellite navigation systems, patient medical trials, patients' diaries, security, asset tracking, dynamic signage, vending machines, stock control, pest control, remote measurement of water and energy usage, automatic meter reading, wireless alarm systems, emergency lighting and wireless closed circuit television apps.

However, the road ahead is not easy. Continuous success for operators in this space will require transformational strategies along with new value propositions, product innovation and perhaps even some mergers and acquisitions. At the same time, operators need to have good insight into the software and system solutions that are emerging as the fastest growing segment in the M2M industry. Increasingly advanced large-scale M2M applications require advanced service enablement platforms that integrate remote devices, mobile networks and enterprise applications.

The challenges involved in exploiting M2M are significant but the potential benefits are even greater. For enterprises that use M2M embedded mobile, there is the potential for greater efficiency, improved business processes, and innovative new business models. For network operators, mobilizing the machine represents a way to greatly extend the subscriber base and drive up data traffic. For cellular device vendors, machines are a vast new target market. For OEMs, embedding mobile in their products allows differentiation, has the potential to expand their customer base, and could even enable new product lines.

--♠--

Telco Challenge # 9 : The Quest for Financial Acumen

The release of additional spectrum is often used as a vehicle for introducing additional competition by the Regulators. From the perspective of spectrum regulators, careful spectrum management is required to ensure that sufficient spectrum is available to support not just the development of the commercial mobile market, but to support the continued operation of critical services such as Government, utility and Emergency Services that use radio spectrum on a daily basis. For the mobile operators spectrum is a critical resource , notably the ownership of lower band spectrum without which 4G will remain a myth in some emerging countries.

As we all know frequencies in the low band range 700MHz to 2.6GHz provide the optimal combination of propagation or coverage (the lower the frequency the better the coverage) and the ability to carry information or traffic (the higher the frequency the greater the data carrying capacity). Indeed some dense urban cities networks are now approaching the limit of network densification and additional spectrum and or new technologies may be the only route for alleviating network capacity constraints. Not to mention to stay in existence over the long haul !!

Unfortunately many mobile markets are no longer experiencing revenue growth as mobile broadband revenues simply offset declining voice revenues. To exacerbate the situation only smart Telcos have figured out how to monetise LTE.Telcos will need to consider the requirements of these air interface choices – such as the levels of handset/terminal take-up, as well as base station, antenna and transmission upgrades – when formulating their spectrum acquisition plans. In emerging countries the demand for high speed data services and delays in availability of Digital Dividend Spectrum has caused severe congestion on their current 3 G networks.

Understanding the value of spectrum to a business is essential for developing a spectrum strategy and participating in a spectrum auction as rightly pointed out by the Coleago experts. The massive cash outlay for additional spectrum and the requirements to make a return on spectrum investments adds a layer of complexity to the evolving cost of data on HSPA and LTE networks. The substitutional nature of some spectrum bands requires a holistic approach to re-farmed 900 and 1800MHz spectrum strategy and valuation. The valuation process must consider stand-alone regional and / or block valuations and also packages of regions and / or blocks. When considering packages over stand-alone valuations the impact of scale must be included.

In respect of a spectrum auction an operator has to find an answer to three fundamental questions:How much spectrum do we need in different bands? The question relates to an assessment of spectrum need in the context of the growth in demand, notably mobile broadband. This needs to take account of the overall strategy, for example traffic offload through WiFi or Femto cells.

How much is each block worth, i.e. what is the most we should bid for it? This relates to valuing each spectrum block in order to set the bid limit for the auction. This is quite separate from auction strategy. Clearly, if there is no bid limit, the auction will be simple because a bidder would simply pay whatever it takes to win the spectrum. However, such an approach may not result in the creation of shareholder value and may draw criticism from shareholders and the financial press and capital markets.

How do we obtain the spectrum as cheaply as possible? In any auction, the bid limits should be set before the start of auction. The role of bid strategy is to ensure the spectrum is obtained for less than the bid limit and at the lowest possible price. Depending on the auction format there may be an opportunity to influence the outcome and avoid negative effects such as aggregation risk (e.g. be stranded with unwanted blocks). This is addressed by examining the auction rules and developing a bid strategy which will be tested through simulations and mock auctions.

Operators may also have to consider mitigating strategies for a "no spectrum case" but if they face network congestion a range of mitigating strategies such as traffic shaping and fixed line off-load must be examined and incorporated into the valuation process.Operators must consider how regulation on net neutrality might impact their ability to shape traffic profiles and whether there are any long run cost implications of offloading to other players fixed networks.

Spectrum is often awarded through an auction process and recent auction designs favour a second price rule which means that bidders cannot influence the price they pay for the spectrum only the price that others pay. A bidder's valuation for a spectrum block is the price at which he walks away from a take-it-or-leave offer. Where aggregation risk is present valuations should be defined over packages, not just individual blocks. A valuation is conditional on information known at the time.

Depending on the auction format there may be dominant bid strategies or ways to avoid negative outcomes in cases where there is aggregation risk. This can be explored at theoretical level, through simulations and mock auctions. In theory a bidder enters the auction well prepared and the auction itself is a mechanical exercise. However, as the auction unfolds there will invariably be some learning which needs to be processed at the end of each day in order to be prepared for the next day's bidding.

The next big Spectrum " land grab " will take place in Africa (the perennial laggard in the broadband era) even as the Regulators dither and delay in the implementation of the Digital Switchover / Dividend. We predict that LTE will really come of age in Africa in 2015 by which time the 700 : 800 mhz and 2-6 Ghz spectrum becomes available thru auctions or beauty contests. At a joint ITU (International Telecoms Union) and ATU (African Telcom Uniion) meeting the outcome saw Africa become the first region in the world to be in a position in 2015 to cohesively and harmoniously allocate bandwidth freed up by the transition to digital television—the so-called 'digital dividend'— to mobile services in both the 700MHz and 800MHz bands.

However if African Regulators dole out slivers of Spectrum to many " wannabe " Telcos in the false notion that this will decrease prices (as they did with Wimax) then expect the same mess : a host of under resourced " Pygmy " operators in each country setting up localised LTE networks with limited coverage and trying desperately to make a decent ROI. This scenario will do precious little to bridge the catastrophic Digital Divide in Africa. In a few years while the rest of the world will be on 5G African Telcos will still be boasting about their " 3 BTS me first 4G network " using their inadequate LTE spectrum allocations.

There is no doubt in my mind that African Regulators will split the LTE spectrum into slivers over many bands. But here is the good nesws : Carrier Aggregation (in LTE A benefits operators with multiple spectrum positions, those with small pieces, and particularly operators that are combining acquired networks. The initial focus is on higher-speed services, but expect more deployments of 5+5MHz carrier aggregation as emerging markets deploy LTE in 2014.By combining blocks of spectrum known as component carriers (CC) , carrier aggregation enables the use of fragmented spectrum and allows LTE-A to meet its IMT-Advanced headline data rate of 1 Gbps. In simple terms bonding Spectrum channels together to create larger channels enables faster wireless services, and reduce opex and capex costs from running multiple networks. Believe it or not LTE A + CA is the 4 G technology for emerging markets.

A Spectrum bid requires a well honed strategy that factors in technical , commercial and financial parameters to balance a subtle equation that underlines 4G data networks. This is followed by bid strategy that will be implemented systematically along a project time line by a " tiger team " drawn from various disciplines. The acquisition of new spectrum and subsequent technology deployment results in massive Capex and Opex.So simply bidding without an all encompassing strategic plan and its flawless execution is a recipe for disaster....even if your Uncle is running the Regulatory Authority !!!

Mergers and acquisitions have been a mainstay of the telecom industry for many years. In the past decade, the telecom industry has spent an astounding USD 1.5 trillion + on M&A activities, investments that have transformed the industry landscape into the competitive playing field we see today. McKinsey reckon that over 80 percent of aggregate deal value has focused on core telecom services and spectrum license market segments, with the remainder targeted toward adjacent markets such as infrastructure, connectivity providers, multimedia and financial services.The twin requirements of nonstop technological advances and ever more capital to pay for them produce an inherently dynamic marketplace that will keep the telecoms industry at the forefront of M;A action around the globe.

"We are clearly looking for distressed assets. If we can find the right opportunity, within specific countries, we will do that, "Big operators buying up whatever they can are rarely a success case. It bites them in the behind later. Operators "should start with in market consolidation, then fixed mobile convergence ... then international – the latter is very complicated" (Mats Granryd, CEO of Tele2)

Wealthy telcos from Middle East and China are on the prowl for well-priced targets both within and outside of their home markets, and in general strategic operators will outbid private equity players for attractive targets. The continued tension between growth in allied fields (for example, telecom operators acquiring cable TV companies or application providers) versus specialisation in one subsector (telecom operators spinning off cell towers) will also serve as kindling for M&A activity throughout the sector.Vodafone agreed to pay 7.7 billion euros for Kabel Deutschland, the country's largest cable company, because it combines phone, Web and TV services to increase customer loyalty and stabilize prices.

In one of the largest M&A deals this year was Telfonica's agreed to buy the E-Plus German wireless unit of Royal KPN NV (KPN) in a cash-and-stock deal valuing the unit at 8.1 billion euros ($10.7 billion) to become the country's biggest mobile-phone operator by customers. The Dutch phone company will get 5 billion euros in cash and a 17.6 percent stake in the combination of E-Plus and Telefonica Deutschland Holding AG (O2D), the Spanish carrier's German unit, which uses the O2 brand. As the EU commissioner in charge of the digital agenda, pushes reform in favor of a single European telecommunications market, carriers have become more emboldened to pursue deals. They are seeking to share expenses to build so-called fourth-generation networks to cope with rising demand for faster data connections

An analysis of the transactional rationale of the Telefonica deal provides valuable insights into the main elements for considerations of " best of breed " Telco M&A deal meaning covering the financial and non financial bases. Ofcourse there is no substitute for old-fashioned focus on the fundamentals of M&A: a clearly articulated and well thought-out strategic rationale for the acquisition becomes the yardstick by which to measure individual decisions that arise during the course of a transaction. Without one, decisions are made that end up being costly and inconsistent with the ultimate strategy chosen – or worse, require divestment of the entire acquisition years later as a 'bad deal'.

First Telefonica's desire to create a Digital Telco Titan : become a leading player with a combined customer base of 43m, 42% in postpaid and derive strong scale benefits with combined mobile revenue market share of 32% . O2 and E-Plus's combined customer base at the end of March would leapfrog Vodafone's 32.4 million and Deutsche Telekom's 37 million, according to data compiled by Bloomberg Industries. Germany has become the hottest battleground for telecommunications assets in Europe as demand for video and music delivered wirelessly and over the Internet increases, while voice revenue declines.

Second motivation was value crystallization through significant synergies. Telefonica is targeting NPV of synergies of €5.0–5.5bn, net of integration costs with projected Net savings from year 2 and Annual run-rate synergies of approx. €800 m; 75% of run-rate synergies by year 4. The deal will result in cost savings and revenue "synergies" of around 5 billion euros. Telfonica identified achievable synergies by rationalisation of distribution network ; increased efficiency in customer service costs leveraging best practices and scale ; better channel management and reduced overheads ; focused rollout on one common nationwide LTE network and improved quality from 3G network consolidation ; backbone, backhaul and core network consolidation, with reduced OpEx from network integration (rentals, power, maintenance, transport costs, overheads) ; site consolidation and rationalisation via   reduction of around 14,000 sites ; increased efficiency by leveraging scalable transmission agreement with Deutsche Telekom and reduced SGA expenses by process rationalisation and a focus on become a more lean agile organisation.

Third motivation Telefonica wants a single LTE network to provide what they call the Best Mobile broadband experience. Key factors included giving customers to benefit from the best high speed mobile and fixed experience from a single LTE network and access to future-proof DT NGA network ; Tariff innovation, voice & video; mobile data bundling ; Strong multi-brand portfolio across segments ; Offering ICT / cloud solutions for business customers ; extensive distribution channel and outstanding customer service ;Leverage convergence through cross-selling / up-selling opportunities as well as profiting from digital innovation and scale from Telefónica's global capabilities (data centres , portfolio of OTT services and partnerships) .

Final motivation was value Creation for Telefónica Deutschland Shareholders . Here they were looking for enhanced financial flexibility (improving leverage) while maintaining an attractive shareholder remuneration ; maintaining conservative pro forma balance sheet with a projected EPS and FCF accretive from first year of full operation. In addition the M&A was all about investing in future growth while reinforcing geographical diversification, increasing exposure to an attractive market with a positive impact on Telefónica's cash flow generation profile.

Telfonica opted for the " Financing Without Increasing Leverage " motto meaning the deal is very positive for Telefonica from a business perspective while it doesn't affect its debt position. They have a Rights Issue in enlarged Telefónica Deutschland of €3.70bn. Telefónica subscribes prorate to its stake of 76.8%, €2.84bn + €1.30bn to KPN for 7.3% stake in the enlarged Telefónica Deutschland. Required total financing of €4.14bn is structured as 50-65% Hybrid, 100% equity under IFRS/ 50% equity for credit rating agencies and 20-30% Mandatory Convertible. Their objective is weighting around 2x incremental OIBDA, excluding synergies ; with Net debt/ratio preserved in short term for neutral to positive impact but keeping strong liquidity to maintain 24 months maturities for FCF generation till deal completion. Economic KPIs and cash flows must be consistent with real value creation. There is no place for speculation, particularly in these variable markets where sources of capital are skeptical, margins becoming tighter, and the consequences of missing forecasts are more direct.

As the folks at Ey rightly point out that a fanatical focus on due diligence of all aspects of the target's business and complete regulatory and market landscape is indispensable when there is so much money at stake . There is no substitute for old-fashioned focus on the fundamentals of M/A: a clearly articulated and well thought-out strategic rationale for the acquisition becomes the yardstick by which to measure individual decisions that arise during the course of a transaction. Without one, decisions are made that end up being costly and inconsistent with the ultimate strategy chosen – or worse, require divestment of the entire acquisition years later as a 'bad deal '.

Proactive Telcos clarify their M/A approach, organization and the way they source deals globally, and work closely with investor relations to ensure they have the right story to tell the capital markets—especially if they aggressively pursue new adjacent areas that have different value creation profiles or emerging market economies with majority low ARPU subscribers .

" We believe M+A as a potential means to transform its business as the industry moves towards software based service delivery. Our previous approach was more focused on filling gaps. We are now looking at a complete view of which new business we need to be in. The performance on acquisition is not necessarily stellar. It's an important caveat. An acquisition is not a panacea, it's not going to solve the problem" (Rima Qureshi, SVP, head of M+A, Ericsson)

Nothing symbolizes the African renaissance better than the mobile phone. It represents technological advancement, deepening connectivity, and economic inclusion. Unfettered by outdated fixed-line infrastructure, Africa is at mobile technology's bleeding edge — pioneering everything from mobile payments to crowd-sourcing. Unfortunately political upheaval and commodity price volatility have posed a big challenge for investors in Africa . Furthermore Telcos are a capital intensive business so there are special ROI challenges in monetising in hyper competitive low ARPU markets that characterise Africa. In any case most Telcos must slim down their various network, sales and service business models, abandon secondary activities and use M+A or cooperative ventures to penetrate fast-growing markets.

Large telcos that operate in emerging markets have higher valuation multiples than their peers in mature markets. Investors place a premium on growth prospects, so long as they are well monitored and diversified . Equity investors we are looking for a business that generates positive free cash flow from wireless and mobile broadband operations by year 5, and a business valuation based on EBITDA positive operation by year 2 and a 5x EBITDA and 2x revenue multiplier valuation .Acquiring a profitable business with a large, established network may lead to future growth, saving time and money that would have been required to build a similar network from scratch. Acquiring a competitor may enable the acquirer to reduce price wars in certain geographical areas, or to protect itself against acquisition by a larger operator.

Capital markets prefer businesses that have a homogenous operational and capital structure because they are easier to understand and value. This is the reason that conglomerates often trade at a discount to the „sum of the parts" and end up being broken up. MNOs essentially have a highly capital intensive infrastructure-based portion of the business (deploying and running the network) and a low capex, innovation-focused services portion. Splitting these two parts could be perceived by investment bankers to be logical as it would enable management to focus, make performance more transparent, and valuation easier.

Cash Returns On Invested Capital (CROIC) is a good measure of company performance because it demonstrates how much cash investors get back on the money they deploy in a business. It removes measures that can be open to interpretation or manipulation such as earnings, depreciation or amortisation. Telcos tend to focus on the existing capital-intensive business (which currently generates CROIC of around 6% for most operators) rather than investing in new business model areas which yield higher returns.

The new business models (monetising Web 2.0 services) require relatively low levels of incremental capital investment so, although they generate lower EBITDA margins than existing services, they can generate substantial CROIC margins .You want to invest in a Telco with a consistent record of sales growth or a Greenfield that can deliver this in their business plan with a committed management team. Mergers in the telecom sector tend to build on existing "triple-play" offers. The emergence of "quadruple-play" offers — bundles of fixed telephony, broadband internet, mobile telephony and TV — are likely to lead to gains in market share and average ARPU, and a reduction in churn rates. There is a class of intelligent network investments that are relatively straightforward to implement and will yield a bigger bang for the buck :

• More efficient network configuration and provisioning

• Strengthen network security to cope with abuse and fraud

• Improve device management (and cooperation with handset manufacturers and content players) to reduce the impact of smartphone burden on the network

• Traffic shaping and DPI which underpins various smart services opportunities such as differentiated pricing based on QoS and Multicast and CDNs which are proven in the fixed world and likely to be equally beneficial in a video-dominated mobile one.

When evaluating wireless investments , the net profit margin is a critical metric : a fat margin means more money to expand operations, refresh network technologies and marketing and brand building . Invested capital is reduced through the deployment of more efficient technologies and processes that enable effective network capacity to be increased (including better network provisioning, traffic shaping, mobile Wi-Fi offload, femto/pico underlay network, network sharing, Multicast and CDN usage etc.Good acqusition targets are Telcos that can rigorously and swiftly farm out everything apart from their true core business can cut costs by as much as 12%, investment spending by 6% and the workforce by 50%. In return, they become more agile, engender a more entrepreneurial spirit and can realize the value of each part of the company.

Telecom-specific assets should be identified and valued using robust valuation techniques and methodologies. Tangible assets are generally valued by applying the cost approach since no prudent investor would pay for an asset more than the cost to recreate it or to reproduce an asset of similar utility (replacement or reproduction cost method) . Bear in mind that Networks are built and then adapted over many years challenge any operator's staff to keep accurate records of exactly what has been deployed. Poor record keeping affects the ability to audit physical infrastructure and connection topology, and also its management and maintenance.This is especially true when there are mergers and acquisitions; not only may there be many inconsistencies in each company's records, but there is also a significant challenge to integration of often disparate management and inventory platforms.

A simple Network assessment and inventory audit (as part of the DUE DILIGENCE process) can reveal various technical parameters in broad network domains such as : Infrastructure components and capabilities , infrastructure topology and traffic patterns , infrastructure design, capacity planning, and scalability , Infrastructure policy services , Infrastructure management and OAPM (Operation, Administration, Provisioning and Maintenance) , Business continuity policies and practices. So you pay for what exists in reality not in a cooked up asset register or a pompous mission statement.

According to Accenture to reflect the real value relationship, the bulk of the due diligence effort needs to focus on helping the acquirer understand the target's future prospects and how those fit with the acquirer's strategy. This can require a disproportionate emphasis on at least two due diligence work streams: strategic and operational. Strategic due diligence involves validating the acquisition target's fit with the acquirer's strategic rationale for the acquisition, and understanding the target's market position and outlook to inform the price offered.

Operational due diligence involves understanding the operational characteristics of the target (for instance, organization structure, IT systems, and culture) and hence the integration approach and timeline that will be required, as well as validating the target's operational and capital expenditure outlook to inform the price offered.In contrast with the financial and legal varieties, high-quality strategic and operational due diligence do not generally require an army of specialist advisors. Rather, the due diligence work streams can be staffed by the acquirer's own people, selectively augmented with advisors who bring targeted insights, such as independent perspectives on the market outlook or an in-depth assessment of the timeline and cost to get to a common IT platforms.

Acquisitions and partnerships are essential for success in emerging market segments such as mobile advertising and cloud computing. However Telcos need to clearly discriminate between when they should acquire and when they should partner. The ability to sustain partnerships will emerge as a strategic differentiator. Effective management and implementation of M+A and partnerships offers significant operational upside to telecom players.In the final analysis a careless approach to investment can be disastrous to the share price is evidenced by the Telkom SA / Multilinks Nigeria debacle. Over time about one billion dollars went down the drain in a classic case of poor INVESTMENT DUE DILIGENCE practices (or lack thereof).

Today's service providers face three critical challenges: getting to market quickly with high-value, converged multimedia services, optimizing infrastructure costs and delivering true carrier-grade performance. To best serve their subscribers and drive new revenues, service providers are transforming their intelligent networks into intuitive networks — that are device-aware, application-aware and access-aware. New differentiating service bundles are being prepared with attractive applications, such as rich voice, gaming, presence, instant messaging, video conferencing and sharing and personalized mobility.

The build out of today's 4G networks such as LTE requires a dramatic increase in computational resources to adequately deliver flexible telecommunications services to mobile subscribers. Yet business conditions also necessitate that new markets are approached incrementally. The challenge for telecom carriers is to reduce the cost of serving the first subscriber in small or cost-sensitive markets. The primary challenge in serving small LTE subscriber bases is that traditional core network architectures require high capital expenditures just to serve the first subscriber.

Networks, whether entry-level or full-scale, are traditionally built using separate network elements for each of several different functions. And most network elements have been deployed with a pair of carrier-grade servers to achieve redundancy with an active and a standby configuration. Thus, a new network with 10 network elements requires 20 servers just to provide service to the first subscriber. Furthermore, because the network is designed to eventually support a large population of subscribers, the servers would remain underutilized until the subscriber base grows to the expected population. The ROI for small and emerging markets has therefore been limited by these high capital outlays. High operating costs for maintaining the servers and providing data center floor space, power, and cooling have also hindered new service opportunities.

The greatest opportunity for revenue growth for wireless broadband presents itself in the form of smaller markets with less than 50,000 subscribers, thereby lowering the cost dramatically to serve the first subscriber and the breakeven point in the Operator's business case . By dramatically lowering the cost to serve the first subscriber, new networks can be built on a campus or targeted community basis with new services tailored to the specific needs of these smaller, targeted markets.

In telecommunication parlance, a "carrier grade" or "carrier class" refers to a system, or a hardware or software component that is extremely reliable, well tested and proven in its capabilities. Carrier grade systems are tested and engineered to meet or exceed "five nines" high availability standards, and provide very fast fault recovery through redundancy (normally less than 50 milliseconds). A rule of thumb is to achieve an availability of five-nines: the system is available 99.999% of the time. This equates to a stringent downtime of 6 seconds in a week or 5 minutes 15 seconds in a year. System availability is dependent on the availability of its components. A chain is only as strong as its weakest link. Thus, if a system needs five-nines availability, then the software should provide six-nines availability and the hardware should provide six-nines availability ($0.999999 \times 0.999999 = 0.999998$).

Scalability is often in reference to architecture. A system that has five units can scale to fifty easily because the architecture allows for it. On the other hand, a system designed specifically for five units cannot scale to fifty because its architecture is inadequate. A modular hardware architecture and decoupled software architecture enable you to deploy IMS services on a very small scale (a single node) or very large scale (a multi-node, high-capacity system). Various dimensions that govern system capacity (such as provisioning, database, transactions and signaling) can all scale separately, so you can apply investments very efficiently.

The global Advanced Telecommunications Computing Architecture (ATCA) standard incorporates the latest advancements in high-speed interconnect technologies, next-generation processors and platform management capabilities. Computer equipment built to ATCA standards will work effectively in the network core of a wireline, wireless or cable provider. Service providers reap the benefits in faster time to market, lower costs and accelerated pace of innovation to introduce new features and services.A typical Service Engine forms the core of platforms built on the ATCA standard to incorporate hardware redundancy, a fault-tolerant software architecture and self-monitoring/ self-healing features. There are separate cooling zones for redundant components, separation of switch hubs to prevent accidental removal or damage, and enhanced fault detection and handling.

The software used on ATCA products is enhanced to incorporate improved reliability mechanisms such as self-stabilizing and fault tolerance. Self-stabilizing software means that the system will more readily converge to an error-free state autonomously. This can be achieved through higher coverage of hardware and software faults, an approach that is derived from Failure Mode, Effects, and Criticality Analysis (FMECA) military standards.

The message for service providers is clear : selected IP platforms are ready to deliver the reliability and availability necessary for real-time, multimedia-rich content, including voice. In an all IP world , multi-core processors coupled with powerful virtualization technology enables the consolidation of all the physically discrete carrier-grade servers into a very attractive platform for low-end scalability. Replacing 20+ carrier-grade servers with either 2 blades or 2 carrier-grade servers based on multi-core processors represents a dramatic way to lower the cost of the core network elements required to serve the first subscriber; this type of radical consolidation represents at least a 10:1 reduction in initial CapEx, plus a comparable reduction in recurring operating expenses.

Forward thinking Telcos must capitalize on the advantages of IP for converging voice, data, and multimedia services on a single unified, cost-effective core infrastructure running on ATCA carrier grade blades. The maturity of IP standards and quality of service (QoS) on IP networks opens up new possibilities for carrier applications. Converging voice and data services over a single IP backbone (such as LTE) maximizes network efficiency, streamlines the network architecture, reduces capital and operating costs, and opens up new service opportunities.

So why is the financial and investment community leery of business plans presented by broadband operators especially when it comes to broadband data ? To start with the Operator tries to bamboozle the financiers with tech speak " Guys we have an all-IP architecture, spectral efficiency OFDMA , bandwidth flexibility backboned on metro ethernet (Yawn) . Since the financers feel uncomfortable (if not downright ignorant) with the aforementioned techno blast they try to baffle the Operator with sublime finance speak " Guys , as equity investors we are looking for a business case that generates positive free cash flow from wireless and mobile broadband operations by year 5, and a business valuation based on EBITDA positive operation by year 2 and a 5x EBITDA and 2x revenue multiplier valuation" (Yawn).

After thoroughly confusing each other with their subject matter expertise both parties scramble to find some common ground : Clarion call " lets BRIDGE THE DIGITAL DIVIDE : meaning connect the poor sods in rural areas who don't even have water or electricity. Atleast thinking about the poor while we gorge on caviar sushi makes us feel human and that can't be bad nutrionally or spiritually that is !! As we all know the investment model for broadband wired and wireless installations must consider all aspects of design, deployment, and integration from the core through the systems architecture, service edge, access network and device. While the initial spend on deployment will have a large focus on capital components associated with procuring the necessary equipment throughout the network and systems architecture, as the broadband network service is introduced and subscriber adoption and usage rates grow, the ongoing operating expenses will consume a growing share of the total cost of ownership.

There are a lot of uncertainties connected to the forecasts.Since the broadband forecasts are developed thru qualitative and quantitative information,statistical modelling and also subjective input to the modelling, it is difficult to express the uncertainty by a pure statistical model. However, it is important to analyse the impact of the broadband forecasting uncertainty.The long-term forecasts are mainly used as input for rollout decisions of different broadband technologies and for establishing new network platforms. Techno-economic assessments are used to calculate net present value, internal rate of return and pay back period for the various projects. A relevant method for evaluating forecast uncertainty is to apply a risk analysis.

Typically the Operators grossly overestimate the market demand and thumb suck the number of new subscribers they will get in year 1 : guys we are targeting 5 % of the total available market without the foggiest clue what the real demand is. While statistical analysis and models are the tools of trade , the sane forecaster has to stand back and take a broader view. If a forecast seems implausible, this is generally because it is implausible. If a forecast results in an extremely high ROI, it is likely that the forecast underpinning the business plan is unrealistic. A very profitable industry attracts more competitors, leading to a loss in market share and increased price competition. This would change the firm's demand and revenue forecast.

A well researched and clearly structured methodology which is based on accepted economic theory and market models instills confidence in decision makers, investors and lenders. While forecasting subscribers and revenues for mobile broadband and specific applications is key to analyze the top line of any business case, translating these forecasts into traffic and bandwidth forecasts is required in order to effectively plan the network and analyze the impact on the bottom line. The moment you get the top line wrong then the whole excel spreadsheet is worthless.

To create accurate market-demand projections for broadband services and assess the availability of alternative technical infrastructures requires testing existing projections against different points of reference, adding context and detail where necessary. Where projections do not exist, we need to create them from scratch with management input. That entails conducting a macro-economic analyses to input to the overall expected level of spend on broadband data ; benchmarking with usage in comparable countries assists with the development of models of how that spend might be broken down by service type, and over time more advanced markets (such as Japan and South Korea) are analysed to assess how mobile data consumption might be expected to evolve over time.

It good to review brokers reports, market research and press articles relevant to the candidate market with care ; we must undertake interviews with relevant experts and panels in order to establish a consumption model for mobile data - where the consumer is likely to be, and what they are likely to be doing when they consume mobile data. To assess opportunities to achieve growth through the introduction of mobile broadband services, we must gather quantitative information and high-level financial information to develop a short business case along the following imperatives :

• Study the competitive landscape, the regulatory background and the alternative technologies available to develop a business model and a strategic positioning reflecting the brand, the skills and the ambition of the operator

• Develop a number of strategic scenarios, consistent with the operator's strategy, and propose models to assess the likely financial impact of pursuing each option

• Include the evaluation of opportunities such a diversification through organic or inorganic growth, and review of the synergies with the existing operator's activities and network

• Construct market demand projections for a comprehensive set of different content types .The output of this stage is a detailed spreadsheet model including projections of spend and usage for each different content type

• Once usage projections have been created, consider the 'optimum' technical support infrastructure for each service type. The result is a summary of the optimal technological choices in a given market, given the expected user profile

• Include the optimisation of returns and the minimisation of risks, as well as the identification and development of strategies for non-conventional revenue streams

Operators planning investment into broadband installations need to be certain that their front-end strategy and planning efforts consider the end-to-end proposition of the network, systems, and service to truly reap the cost benefits and the revenue potential of broadband wired and wireless services. And then and only then you might get lucky with the investment banking community. Long-term broadband technology forecasting is not a very easy subject. Experience has shown that it is nearly impossible to make long-term forecasts without understanding the evolution of new broadband technologies and new broadband network platforms. Knowledge of broadband technologies regarding possibilities and limitations is important for the forecasting.

In order to make good long-term broadband forecasts,techno-economic analysis of the relevant broadband technologies has to be performed.Each technology generates investments and operations and maintenance costs for the rollout, which is dependent on the characteristics of the various access areas in the countries. The techno-economic calculations evaluate the "economic value", i.e. expressed by net present value or pay back period of rollout of different broadband technologies.The assessments are carried out for rollout on a national level and on specific areas like urban, suburban, rural and especially the rest market to examine the potential of the different broadband technologies. Therefore, the techno-economic analysis is crucial for technology rollout strategies and for broadband forecasts.

Wireless networks combined with submarine cable bandwidth will offer faster Internet connectivity and advanced multimedia communications. Location awareness opens a fresh business opportunity for the Telco value chain as it uses a person's location to provide compelling services and experiences. Smartphones and Tablet PC's will leverage Cloud Computing solutions to enhance enterprise productivity. M2M market (connecting devices other than phones, laptops and similar consumer devices) is the new gold mine for network operators seeking data revenue

---♠---

POST SCRIPTUM : A peek into the Future

5G : The Black Swan has landed

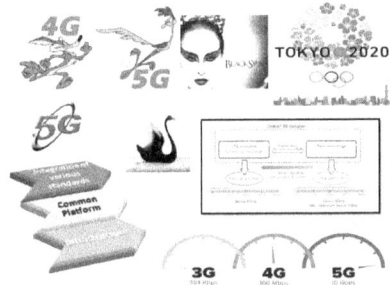

Despite 5G technology not yet being standardised and unlikely to be ready for half a decade, many in the industry are breathless with excitement . While vendors jockey for mindshare, there is no shortage of posturing around "what 5G is" and who should care. This includes the technologies that will make 5G a reality (including virtualisation, millimetre wave spectrum, unlicensed spectrum, duplex free operations) along with 5G drivers like ubiquitous IoT demands. Next Generation Mobile Networks Alliance feels that 5G should be rolled out by 2020 to meet business and consumer demands.

So what are the characteristics of a 5G network ?? For starters it is a super-efficient mobile network that delivers a better performing network for lower investment cost. It addresses the mobile network operators pressing need to see the unit cost of data transport falling at roughly the same rate as the volume of data demand is rising. It would be a leap forward in efficiency based on the IET Demand Attentive Network (DAN) philosophy. 5G is super-fast mobile network comprising the next generation of small cells densely clustered together to give a contiguous coverage over at least urban areas and gets the world to the final frontier for true "wide area mobility". It would require access to spectrum under 4 GHz perhaps via the world's first global implementation of Dynamic Spectrum Access. And yes it is a converged fiber-wireless network that uses, for the first time for wireless Internet access, the millimeter wave bands (20 – 60 GHz) so as to allow very wide bandwidth radio channels able to support data access speeds of up to 10 Gbit/s.

In order to further extend traffic capacity and to enable the transmission bandwidths needed to support very high data rates, 5G will extend the range of frequencies used for mobile communication. This includes new spectrum below 6GHz, expected to be allocated for mobile communication at the World Radio Conference (WRC) 2015, as well as spectrum in higher frequency bands, expected to be on the agenda for WRC 2019.LTE will continue to develop in a backwards-compatible way and will be an important part of the 5G wireless-access solution for frequency bands below 6GHz. Around 2020, there will be massive deployments of LTE providing services to an enormous number of devices in these bands.

While both 3G and 4G radio access networks (RANs) were built as stand-alone network, 5G RAN will be deployed by integrating the existing LTE-Advanced (LTEA), its evolution technologies, and new radio access technologies (RATs). So expect Virtualization and SDN (the other big thing in network evolution) will be extended to 5G mobile wireless networks as well. As you know with wireless network virtualization, network infrastructure can be decoupled from the services that it provides, where differentiated services can coexist on the same infrastructure, maximizing its utilization.

In an idealised 5G core network the control plane is separated from the data plane and implemented in a virtualized environment. It is a fully distributed network architecture with single level of hierarchy built to provide shortest possible delay in data path. Developing an SDN business involves the deployment of physical infrastructure, a network controller and a telecoms operating management system which combines operation and business support systems. The network controller is central to SDN with two main functions: virtual resource control and traffic management systems (TMS). The network controller can create a programmable, logical network that allocates resources within the physical network (access and core networks) in the most dynamic way without needing to know the actual infrastructure topology. In so doing, the operator can build the most appropriate virtual network offering multiple services.

Carriers that have jumped headlong into the SDN/NFV bandwagon have positioned themselves (knowlingly or unknowingly) to launch 5G !! Telecom Italia has been among the tier 1 telcos driving the move to NFV. Along with AT&T, BT Group, Deutsche Telekom, Orange, Telefonica and Verizon, the company a couple years ago pushed network functions virtualization into the spotlight by creating an ETSI group to explore the technology.

Meanwhile AT&T, has introduced its vision for the company's network of the future: the 'User- Defined Network Cloud.' ATT claims their the cloud-based architecture is "a global first at this scale."The carrier expects its revamped architecture will accelerate time-to-market for technologically advanced products and services. Integrated through AT&T's wide-area network (WAN) and using NFV and SDN, the architecture is expected to simplify and scale AT&T's network by separating hardware and software functionality, separating network control plane and forwarding planes, and improving functionality management in the software layer.Of course Telco Industry Execs are waxing lyric (and not without just cause) about 5 G and its titanic potential. Here are some comments.

Middle East operator Etisalat claims the rollout of 5G technology is one of its goals for 2020 and will underpin its future support for machine-to machine (M2M) and eGovernment services as well as the wider Internet of Things (IoT).The Internet of Things (IoT) has become shorthand for the concept of billions of connected devices. And it's an idea that's getting some real traction as big names position themselves as players in the market.

" The mobile industry should consider 5G as a "special generation", introducing challenges in all layers of the technology" (Mike Short, VP of public affairs Telefónica Europe)

" We have engineering teams working on LTE and 5G. Each time the 5G team unveil a new performance leap, the LTE engineers respond by matching it " (Matt Grob, CTO for Qualcomm)

"5G is a fundamental change in technology and will have a significant impact on how we offer services. We must look at performance and coverage, and not just consider microcells." (Allan Kock, director of RAN development at TeliaSonera)

" 5G must have green as part of its very DNA, ensuring that all aspects of 5G, from access networks, data centres and transport network to connected devices only consume energy when they are being used. However 4G is a success, let's enjoy . We shouldn't jump too fast," " Stephane Richard Orange France

"Most of our competitors talk about 5Gb/s and 10Gb/s or some other number, but they're not telling you the configuration, which is ridiculous. 5G should not be a "forced leap" in technology, but draw heavily on re-use existing radio air interfaces – LTE, WiFi and LTE-U – under a common control plane." (Marcus Weldon, CTO of Alcatel-Lucent)

"Computing costs have fallen 1000-fold since their inception.We've achieved this change with semiconductors, we now have to do the same with 5G compared to 2G," KT Telecom CEO Chang-Gyu Hwang

Hans Vestberg, President and CEO, Ericsson, says: "More and more industries tap into the value of digitalization and connectivity. 5G will amplify that as it is designed to be the industrial internet. It will not only be built for consumers, but also for digitalization of industries and the Internet of Things. Together with TeliaSonera we launched the first commercial 4G network in 2009, we will be in the forefront of 5G as well."

Potential 5G applications could include e-health with real time surveillance of patients and remote treatment; connected cars including critical communication between vehicles (warnings, support to self-driving cars etc.) as well as better network performance in terms of capacity, coverage and power consumption.

So where can you go to see 5 G in action ? How about scheduling a trip to the 2020 Olympic Games in Japan.The Japanese government is aiming to put the world's first fifth-generation communications system for mobile phones into practical use by 2020, in time to help promote Japanese technology at the Tokyo Olympics and Paralympics. The 5G system will offer a communications speed about 100 times faster than LTE. For example, when a user downloads a two-hour movie online, it takes about five minutes with LTE, while a 5G user would be finished in several seconds. Users will also be able to enjoy movies with high-resolution such as 4K on smartphone and other portable terminals.

As for how we watch the games, Japan's national broadcaster Nippon Hoso Kyokai is at work on 8K Super Hi-Vision broadcast quality that offer 16 times the level of detail of current top-of-the-range 4K systems. NHK aims to begin transmission in the 8K system by 2016, and to be able to broadcast the entire games with exceptional quality by 2020.

Johan Dennelind, President and CEO, TeliaSonera, says: "Our ambition is to be at the cutting edge - at all times - offering our customers and society at large all the possibilities that technology brings. Stockholm and Tallinn are two of the most connected cities in the world and now we'll take them to the next level. 5G will create completely new innovations, ecosystems and great services to our customers. 5G will also take connected things (IoT) to a new level. I can't wait to see how Stockholm and Tallinn will embrace 5G."

5 G is actually a " Black Swan " . According to author Nasim Taleb a Black Swan is an event with the following three attributes. First, it is an outlier, as it lies outside the realm of regular expectations, because nothing in the past can convincingly point to its possibility. Second, it carries an extreme 'impact'. Third, in spite of its outlier status, human nature makes us concoct explanations for its occurrence after the fact, making it explainable and predictable.

So believers get ready : 5 G will amplify the challenges to rock your world !!

---THE END---

www.ingramcontent.com/pod-product-compliance
Lightning Source LLC
Chambersburg PA
CBHW070247190526
45169CB00001B/325